Circle of Brightness:
*Rural Americans Recall
The Day The Lights Came On*

*A Project of the George W. Norris Foundation,
McCook, Nebraska*

Edited by Brent L. Cobb

Copyright © 2002
by
George W. Norris Foundation

All rights reserved. No part of this book may be reproduced in any form, except for the inclusion of brief quotations in a review without permission in writing from the author or publisher.

Library of Congress Control Number: 2002095244
ISBN: 0-9725927-0-9

Printed in the United States by
Morris Publishing
3212 East Highway 30
Kearney, NE 68847
1-800-650-7888

Senator George W. Norris

Table of Contents

Preface vi
Chapter 1: *The Dark Ages* 2
Chapter 2: *Early Initiatives* 30
Chapter 3: *The Rural Countryside* 50
Chapter 4: *A Woman's Work* 82
Chapter 5: *Political Shadows* 108
Chapter 6: *Waiting and Wiring* 132
Chapter 7: *Let There Be Light!* 162
Chapter 8: *Technical Difficulties* 182
Chapter 9: *Hallelujah!* 208
Chapter 10: *Relief from Drudgery* 230
Chapter 11: *Enchanted Water* 248
Chapter 12: *Social Illumination* 264
Chapter 13: *Special Days* 288
Chapter 14: *Yesterdays Today* 312
Bibliography 342
Acknowledgments 344

Preface

The George W. Norris Foundation, in an effort to promote the legacy of Senator Norris, sponsored a nationwide essay contest asking rural Americans to write their recollections about the day their rural homes received electricity. Writers from 17 states responded. The Foundation was inspired with the wide array of stories recalling almost every aspect of the rural electrification process — from the behind-the-scenes politics to touching family memories. This collection of stories offered a unique opportunity.

While there have been a number of scholars who have published historical and political analyses of the rural electrification process and about George Norris's career, the Foundation decided to offer another perspective — the people's perspective. It seemed natural since Norris was known as a man of the people since he was the principal author of legislation creating the Tennessee Valley Authority in 1933 and is known as the legislative "father" or rural electrification.

George Norris believed the mass of mankind was good and that people would triumph despite the barriers placed before them. This book is about the triumphs of both George Norris and the American people he represented.

In his autobiography *Fighting Liberal*, Norris said. "In the way our life has been molded, there is a spirit in all of us that resents injustice. We want to see honest service rewarded. We demand that ability and loyalty be recognized. We accord our respect, our admiration, and our love to those millions of Americans who live quietly and simply, without pretense, envy, malice, or ill will."

It was Norris's respect, admiration and love of those Americans that gave a voice to those quiet Americans — most notably those who lived simply and made their living off the nation's rural countryside.

This book is the opportunity for those rural men and women to tell what rural electrification meant to them. This is their opportunity to honor Sen. Norris and countless others who worked hard to bring electricity to the farms.

There are many testaments in this book but perhaps Roye Lindsay's story (Chapter 14) best sums up the the significance of rural electrification. Of a home built in 1887, he writes of births, deaths, wars and Christmas, but "flipping the switch to turn on the electricity to that house affected my life and the history of my family more than any other event."

Several letters prefaced their stories by apologizing for sloppy hand writing citing various physical ailments ranging from arthritis, degenerative blindness, Parkinson's, and stroke. Yet, these people took the time and in some cases — a great deal of effort — to put their stories into words. Many of these writers share the similar experiences but grew up feeling isolated from the world.

Since the inspiration of this book came from the people, most of its elements were created with the people in mind. The book's title, "Circle of Brightness" is taken from a phrase used by one of the writers, Gay Price (Chapter 7) who describes having to read or play games close to the kerosene lamp and its circle of brightness. Until George Norris and others led the way for rural electrification, rural people's lives were confined, on many levels, to that circle of brightness.

Chapter titles are also derived from the words and phrases used and written by the people. Everyone who felt rural electrification important enough to write an essay is included. Every editorial effort was made to keep the stories in the people's voice.

In the people's words, these are stories of laughter and tears, of hope, despair, fear, joy and ultimately triumph.

— Brent L. Cobb

*This book is dedicated
to those who worked at every level
and continue to work
to make rural electrification a reality.*

Chapter 1

The Dark Ages

Kerosene lantern maintenance was a daily ritual.

Norris: A man of the people

As one decade eroded into the next, electricity energized the people of the cities with a future of growth and prosperity. Yet, for a nation's rural countryside, life existed much the same as it did for their pioneer ancestors. Unless something changed, rural Americans would fall farther behind their city cousins.

Farm families needed a voice. They needed someone to unify their push for rural electrification. They needed someone who understood milking cows by lantern light, and heating water on the stove for Saturday night baths. They also needed someone willing to come up with solutions to economic concerns, answers to political questions and practical solutions. They needed a farmer to fight for the nation's farmers. They needed a gentle man with the right amount of patience, courage, perseverance and initiative.

They found their man in Nebraska senator George W. Norris, and he was one of them.

He was born in the gently sloping Ohio timber country east of the Sandusky River and south of Lake Erie's Sandusky Bay. His father cleared 25 acres of trees and huge stones and made a frontier home with his wife and 10 children. The youngest of those Norris children, George William, came to know seven United States Presidents and became known as the father of rural electrification.

In 1864, before he turned four, George's oldest sibling and only brother, John Henry, was shot and died while fighting for a Union victory at Resaca. Seven months later, George's father died from pneumonia a few days after a spooked team of wagon horses bolted and threw him to the ground.

Six of the nine girls remained at home, and George's 46-year-old mother was six months pregnant.

The farm still needed worked and George watched his

mother tend the garden, plant crops, and assume the family's financial planning, as well as continuing her duties of spinning the jenny, weaving, sewing, washing, ironing, canning, and nurturing and encouraging her children and always striving for a better way of life.

From these humble beginnings, Norris learned much and never strayed. He was elected to the U.S. House of Representatives in 1902, served five terms and was then elected to the Senate in 1912 and served five terms through 1942 but he never forgot from where he came.

What Others Said About George Norris: On July 11, 1861, when Sarah Melissa Norris hurried about the hills of Northern Ohio to spread word that her mother delivered a new baby brother George — the 10th Norris child — a neighbor lady exclaimed:

"Sakes Alive! What does she want with another? Hasn't she enough work as it is?"

In George's Words: "My father and mother met at a house-raising — that good neighborly ceremonial of an early America, at which people gathered and pooled their labor to build homes for youths and maidens shortly to be married."

In the words of the people, here are their stories about "The Dark Ages" . . . .

Pre-electric days rough

by Ogreta F. Thomas
Allendale, South Carolina

Life was hard in our homes before electricity. We lived in a four-room house out in the country, with an outdoor privy. Women cooked on stoves heated with wood and often had to chop the wood before cooking with it — if the men were busy on the farm. During canning season, the stove would be hot all day and into the night because everything that was grown had to be canned for the winter months.

Water had to be drawn from a well in the back yard. Ours had a bucket on each end of the rope. Animals had to be watered and washed down. Wash day was just that — it took all day to wash and dry for a large family.

Bathing was even a chore. We'd draw water and let it heat in the summer sun, often several children bathed in the same tub. I can't figure how the last ones could say that they were clean.

> "... often several children bathed in the same tub. I can't figure how the last ones could say that they were clean."

Ironing with flat irons was pure drudgery. We'd build a fire around the wash pot to heat the irons and iron in the shade. That was another day's job.

At night, we'd make pallets on the floor near a door to try to sleep when the nights were hot and muggy.

Studying or reading was rough, and most of it was done around the kitchen table. One child was responsible to be certain that the lamps were clean before dark. Children were taught obedience then, so everyone pulled together to accomplish the almost impossible tasks.

Our house had been wired with one electrical dropcord in the center of each room. When electricity came to our home in the late 1940s or early 1950s, we must have felt like Noah when he saw dry land — the day the lights came on was heavenly, astounding, unbelievable, wonderful. And we were thankful.

Our children were finishing their studying when . . . BINGO! There was light!

We laughed. We cried. We hugged and we prayed. Each child had to cut it on and cut it off and then it was my turn. Smiles were from ear to ear. I couldn't sleep that night. I read everything in the house — maybe twice. The next day I ironed and ironed, maybe everything more than once. My husband bought an electric iron the day they started surveying. We all thanked God for his great blessing and prayed that electricity would never be cut off.

Circle of Brightness

Grasshopper memories

by Adeline Bechtold
Watertown, South Dakota

I was born in a sod house and lived in a sod house during the Depression. My dad had a stroke and I have faint recollection of him. We took him in our Model-A to a train in Utica and he later died in a hospital in Yankton, S.D.

During the Depression, the sky was hazy with dust and usually windy and hot. We wore straw hats to make things easier. I was always told to wear my hat so I wouldn't get sun stroke.

The days were hot and sometimes it seemed like there would be a tornado. We took our dog and lantern and climbed into our cellar. The wind blew, but again there was no rain.

The grasshoppers were so numerous they even got into the house and ate our clothing. We had a small building for chickens that could be pulled into a field, whereby the chickens would eat the grasshoppers and we then ate the chickens. I stepped on grasshoppers that were laying eggs on the road all the way home from school. Rabbits were also plentiful and my brothers would hunt them with the Model A at night and come home with a pile of them. Later the rabbits developed tumors, so we stopped eating them.

My clothes were made out of chicken mash sacks and flour sacks. Two sacks made a gathered shirt, one made a blouse.

> "The grasshoppers were so numerous they even got into the house and ate our clothing."

Horses pulled a sleigh, as the roads were not open during the winter. Sometimes the horses had to be shoveled out to make progress. Some of the horses died of sleeping sickness.

Many people moved from our state to California, where there was not a drought. Some just left with motorcycles.

My mother could not understand why some families had so many children when they could not clothe or feed them. Some of the pupils came to our school without coats or hats. We did receive some surplus from the government at our school, which was apples, oranges and grapefruit.

My oldest brother was a talented person. He repaired all our machinery and also was my doctor. My ring finger was infected and he lanced it with a blade and squeezed out the pus. I almost lost my life several times. At an early age, I fell out of a highchair and turned blue. Another time, I had pneumonia and was given a live-or-die medication. Once, a cow trampled me and broke my leg.

The kerosene lamps, gas lamps and Aladdin® lamps had to be maintained. Later we got a wind charger that gave us lights at night. On a windy day you could even iron. Those batteries would bubble in the attic.

The sod house was comfortable and cool and I'm glad that I lived in it in the 1930s.

The meaning of comfort

by Robert H. Currie
Cornell, Wisconsin

The first and most uncomfortable situation that comes to my mind is the outhouse. What an adventure when the weather was below zero! Some people were more ingenious than others — they had a removable seat that was kept in the house and taken back and forth so it would be somewhat warm when you sat on it.

What about water? We had to go outside and pump water by hand and carry it into the house. Be careful when it is cold because your skin would stick to the pump handle.

What about the hot water? Water was heated on a wood burning stove for washing dishes, clothes and bath water. Clothes were washed in a No. 2 tub with a washboard and a bar of soap. On Saturday night, the washtub became a bath tub. Clothes were ironed by heating flat irons on the wood stove.

The house was heated the best it could be with a wood or coal burner called a parlor furnace. Our house was a two-story house, and the only heat we had upstairs was what drifted up, before it drifted out of the house. Where the wood came from, is another story, but it didn't come easy.

Our entertainment, during this cold weather, consisted of listening to a radio that operated on a six-volt battery. This had to be taken up town to be re-charged when the charge was used up.

Of course there wasn't any kitchen utensils such as refrigerators. Not even electric coffee pots or toasters. Lights were kerosene lamps and lanterns, and the light was poor at its best. We almost lost our eyesight doing home work with such poor lighting.

We didn't have a car when we were growing up but

those that did, didn't have an engine heater or battery charger to help get them started in cold weather. Some would take coals from the stove, put them in a pan and slide them under the car to heat the engine. This caused some fires from gas and oil slipping from the engine onto the hot coals.

My family did not farm but we had relatives who did and all their milking had to be done by hand with light provided by lanterns hung up in the barn. The milk was cooled in cans, placed in tank water.

Of course, there was no phone service, TV, microwave ovens, computers or any other conveniences provided by electricity.

Electricity made all of those conveniences possible.

The most comforting change was from the outhouse to an indoor bathroom. Now it didn't make any difference what the temperature was, you could be comfortable.

> "The most comforting change was from the outhouse to an indoor bathroom."

We could have running water in the house, heated in a water heater. The wash tub was replaced with a washing machine and the bath tub moved to the bathroom.

Modern kitchen and household furnishings were made possible. Farming became much easier with the availability of electricity. Heating and lighting systems became more efficient.

All in all, life became more enjoyable and comfortable with the changes made possible by electricity.

Not the 'Good Ol' Days'

by Caryl Kralik
Newman Grove, Nebraska

I remember the days before the REA. To me it was not the "Good Ol' Days." At night, we used candles, lanterns, kerosene lamps and a gas lamp for light. It was difficult to read and sew and there was always a danger of fire.

We were lucky in the 1930s to have 32-volt electricity, so we had a radio and lights. But when we were wired for REA it was so much better, since we didn't have to run the motor anymore and charge the batteries for the 32-volt system. We could have an electric motor on the washing machine instead of gas. Later came the automatic washers and dryers which really changed wash day. Electric toasters and mixers, plus all the appliances available now, really helped the housewife.

Mother heated flat irons on the stove to iron our clothes. Later she had a gas iron and finally an electric one. We swept our carpets with a broom or hung them on the clothes line to beat the dust out. It was great to get electric vacuum cleaners.

Now we didn't have to go to the hog pen and pick up cobs which were burned in the cook stove. What a smell that was! The ashes had to be carried out from wood and coal stoves. It was much better to have an electric range and oven as the temperature could be kept even and there was no mess.

We raised a lot of chickens and had electric brooder stoves to keep the baby chicks warm. The setting hens didn't need to hatch the eggs and keep the chicks warm anymore. Heat lamps could now be used to warm baby pigs, calves and lambs. Electric fences now keep the livestock in their pens. We never milked much and did ours by hand, but those that milked a lot were glad to have electric milking machines.

Electricity changed our lives for the better!

'Hard way' was 'only way'

by Mamie Lee Brown
Kershaw, South Carolina

Before electricity came on, life was hard. Everything was done the hard way. To have water for a bath, someone had to go outside and carry water in from a spring. Clothes had to be washed in washpots. No one complained, though, because electricity was not around. Another way of life was unknown.

Every day was rough living before electricity came on. The chores assigned were much harder than just washing the dishes or making the bed. Instead of buying wood, people had to go outside and cut it in order to have a fire or to be able to cook. Cotton had to be picked and corn stalks were cut and knotted.

There were no complaints about the load of labor because that was what people were used to. They were used to having kerosene lamps and outhouses. It was the only way of life they knew.

> **"There were no complaints about the load of labor because that was what people were used to."**

When the lights first came on everyone did not get electricity right away. It took a while before electric cooperatives could install it in certain homes, especially in the country. Electricity was the best thing that could have happened.

The day the lights came on changed the lifestyles of everyone.

Circle of Brightness

Measuring electricity's impact

by Anna M. Sanda
Williston, North Dakota

How can we measure the impact of the rural electric program? We can compare, but it's hard to measure.

I remember keeping meat frozen in a tin box on the north side of the house during the winter. We locked it to protect it from animals, and we frequently found paw prints in the snow around the box in the morning. When summer arrived, we kept only enough meat on hand for two or three days.

We carried water from the windmill for the house — and contrary to popular belief, the wind does not blow every day in North Dakota. I would sometimes dash back and forth to the windmill several times with the water pail before I got it filled. That great wheel would often quit turning when it saw me coming.

I had a running battle with our gasoline-powered washing machine. It started for my husband Carl, but not for me, no matter how I tried. But Carl would step on the pedal and it would immediately roar to life. And I'll swear as soon as Carl left the yard, that machine would stop, it always knew when he was not around. That would leave me standing helpless, my washing unfinished and the water cooling in the tub.

The Aladdin® lamp was a thing of beauty. Its tall, graceful glass chimney and shiny base looked elegant on the table. But I would come into a slowly darkening living room to find my husband moving his chair closer and closer to the lamp, trying to read the newspaper as the chimney gradually blackened with soot.

We took those things for granted at the time. Carrying

oil into the house every day for the oil-burning heater, warming water for a sponge bath, breaking the ice on the water pail in the kitchen on frosty mornings were a way of life.

And let's not forget the outhouse. Sometimes when I enter a warm, well-lighted bathroom, I think about the chilly spring day I encountered a skunk in the outhouse — or perhaps I should say the skunk encountered me. I dashed for the house, throwing my clothes off as I ran. By the time I hit the back step, I was down to my underwear.

> **"I think about the chilly spring day I encountered a skunk in the outhouse — or perhaps I should say the skunk encountered me."**

But everything changed with the coming of electricity in our area in 1951 — the refrigerator and freezer, hot and cold running water, electric lights, my husband's power tools and electric motors.

Yes, we can compare life then with the way it is now. I often do. Especially on wash day, when my beloved automatic washer starts at the touch of a button — my touch no less.

Comparing is fun, but it is hard to measure the great change in our lives the day the lights came on.

Milking by dim lantern light

by Bob Woodruff
Ulysses, Nebraska

The days of milking cows, with only the dim light of the lantern, moving on a wire line across the barn, is quite vivid in my memory.

At one time, during the early 1930s, my father milked 21 cows by hand, twice a day.

Many mornings he was greeted by a bum, who had spent the night in the warm barn. The railroad tracks ran near the house. The barn is down a hill from the house, with the Blue River running between the two.

After the chore of milking, the cans of milk had to be pushed or pulled up the hill, on a cart or sled, in snow, mud, or both.

When the separating of the milk was complete, the skimmed milk was hauled back down the hill, to be fed to young calves and pigs.

The cream was poured into 10 gallon cream cans and shipped by rail, from Ulysses to the Lakeville Creamery, Lakeville, Minn.

The Blue River caused many problems on the Woodruff farm. Flood waters meant moving the cattle and hogs to higher grounds, often times in the dark of the night. Electric lights would have made doing chores so much easier and safer.

In the winter time, besides chores, ice had to be cut from the river and hauled with horses and wagon to the ice house on the farm.

The ice was packed in layers with straw to keep it from thawing, when the weather got warm. It was then put into the ice box to help keep perishable food. What a great day, when the refrigerator made its way into our lives.

Our house was heated with a wood furnace. Putting up wood was a hard and dangerous job. Many times the wood supply in the furnace was burned up by morning. It was hard to climb out of the warm bed to get dressed. Many times the water would be frozen in the drinking pail. How much more efficient our electric heating systems are today.

The battery-pack radio played an informative and entertaining part of our lives.

We had wires stretched across the attic to get better reception. The radio was used conservatively, as the batteries would go dead at the most important times. How we enjoy our electric radio today.

Circle of Brightness

Nothing went to waste

by Phyllis J. Brubaker
Hesperia, Michigan

    We lived two miles from a little country school in Saginaw County, Michigan. The first two years of my school days, I was driven to school by my brothers. They had a pony and a buggy. A neighbor, who lived across the road from the school, suggested that we drive to school and put the pony in his barn during the day. I remember sitting on the floor of the buggy wrapped in an army blanket.

    We washed our clothes on a scrub board and hung them outside. I had to hang them from the longest item to the shortest. In the winter my fingers would freeze. When we took the clothing inside it was frozen stiff and looked like pieces of cardboard. We heated two irons on the cook stove. While you used one, the other was reheating. As the clothes thawed, we ironed them dry.

    I remember sitting around the kitchen table, playing "set back" with the dim light of a kerosene lamp. We had kerosene lanterns to take to the barn to do the evening and morning milking. During the summer months, the lanterns were not needed. We strained the milk, put it into crocks, and put the crocks on the dirt floor in the basement. The next day we skimmed the cream off the milk, and fed the milk to the animals. Once a week we took the cream into town and sold it — no health department regulations.

    We baked our bread. Mother put her hand in the oven to test. If it felt hot enough, we put the loaves in to bake. The bread always baked well. There was a reservoir (a metal box built on the side of the stove with a cover) on most of the wood cook stoves. We put three pails of water in this tank. We had warm water, if we had a fire in the cook stove.

The Dark Ages

My brother would hitch the horses to a sleigh or wagon. He would drive into town and bring back coal. We never had the money to buy very much. At night, we banked the pot belly stove with coal. During the day we used corn cobs and wood.

We pumped the water from the outside well. We always had a pail of water with a dipper hanging on the side. Everyone drank from the dipper, and put the dipper back into the pail. We also pumped water and carried it to the animals.

We took a bath once a week in front of the cook stove in a tin tub. We carried the tub into the kitchen from the outside shed, pumped the water, carried it in, heated it on the cook stove, poured the warm water into the tub, and took a bath. Next, we emptied the tub, carried water outside, and put the tub away. Our neighbors only bathed during the summer, in the creek.

> **"Grandmother said she'd 'can the squeal of the pig if she could catch it.'"**

We had outside johns. We didn't waste time reading in the john. Everyone had a Sears catalog for tissue. We rubbed a page back and forth in our hands to soften the paper.

We canned everything and anything that was available: fruit, wild berries, vegetables and meat. We stored vegetables in the root cellar. Nothing was wasted. My grandmother said she'd "can the squeal of the pig if she could catch it." We also made a salt brine to keep our meat from spoiling. The brine was put in a large crock and meat was put into the crock. Something heavy was put on top of the meat to keep it down into the brine.

When the electrical company got to our area, the first thing my parents purchased was a washing machine. We had 10 children and relatives were always visiting, sometimes for weeks while looking for a job.

Circle of Brightness

Surviving a lifetime of change

by Pearl M. King
Sarles, North Dakota

I'm an 80-year-old farm wife who managed to survive the pre-electricity days.

Can you imagine no T.V, electric typewriters, frozen foods, dishwashers, clothes dryers, electric blankets and air conditioning? You had to haul your water in and heat it for the baths and then carry it out again.

We used to heat our old sad irons on the wood stove in the summer to iron our clothes. Believe me they were sad.

Not having the luxury of a refrigerator or deep freeze, we had to cure our own meat with salt, or else can it. We baked eight to 10 large loaves of bread from scratch, again in the wood stove.

Now that I am older I realize what a blessing electric power is. We had to turn the old cream separator by hand and make our own butter in a barrel churn. It took a hefty arm to turn it, to say nothing about milking machines.

The kids in those days had to make their own fun, with whatever they had. It was a bright spot if the neighbor had a pony! However, I venture to say, kids were just as happy then.

The power came to our farmyard on Christmas Eve 1950. Can you imagine what a merry Christmas that was? Every neighbor had a yard light and after that the country was no longer dark and scary.

I hope the young people realize how good they have things. They can have a couple of showers a day and use laundromats to do their laundry and they think nothing of it. I realize that computers and the Internet are out of my league, but I am thankful I was able to witness so very much change in my lifetime.

Lighting before the REA

by John Zurian
Moquah, Wisconsin

It was in the early 1920s when my family lived on a farm in northern Wisconsin. My dad would get up early on cold, dark mornings and build a fire in the pot belly wood-burning stove in our kitchen, and light a kerosene lantern. He'd put a few strike-a-lite matches in his shirt pocket, put on his boots and sheepskin mackinaw, and go down to the barn.

The barn was about 200 feet away from the house. The matches in his pocket came in handy when the wind would blow out the light in the lantern he was carrying. In the barn he would light another lantern hanging from a rafter. The two lanterns each gave off about 20 candlepower of light if the globes were clean. That was enough light to let you know where you were in the barn.

Later the Aladdin®, one-mantel kerosene lamp, and the Coleman® two-mantel gasoline lamp were introduced. The Aladdin® gave off about 100 candle power and the Coleman® 200 candle power of light.

In our little community several merchants used Delco® light gasoline power plants to light their stores. They were either 32 volts, with backup storage batteries, or a direct 110-volt power plant. These needed attention every evening and it was quite a chore for the keepers.

Since 1936 our farm had a water wheel churning out electricity harnessed from a creek that ran through our land. It was only a six volt battery, but it sure beat kerosene lamps and lanterns.

Circle of Brightness

No excuses, no vacations

by Thelma Clardy
Pickens, South Carolina

Fifty-seven years ago, our family lived and worked a 55-acre farm down a lonely dirt road. Back then, if you had a pair of mules, land, water and wood that was considered a good farm!

We had a smoke-house, an outhouse, a well house, and a barn all located within a safe distance of our house.

In our family, you sawed wood using a crosscut saw with an eight-year-old girl on one end and a big Daddy on the other. Daddy would take an axe and split wood for the cook stove.

A cast iron wood stove used wood in a firebox to heat the oven. The top had four eyes and a tank to heat water. We dipped warm water into dish pans and wash pans. A bucket and dipper were used for drinking water. The water was hand drawn — a bucket full each time — to fill a No. 3 washtub and a black washpot, each holding 10 buckets for bathing and washing clothing. The well had a box, a log, a Winlett, and a pulley. The pulley had a rope with a bucket on the end and we had to wind the bucket up and down the well.

> "You sawed wood using a crosscut saw with an eight-year-old girl on one end and a big Daddy on the other."

The family did things together daily before dark with no excuses and no vacations.

We learned to fill lamps and lanterns with kerosene. Children were field broke at an early age. The cows were milked and the milk was carried to the spring to cool it.

The Dark Ages

Vegetables and fruits were canned or dried while meat was salted and hung in the smokehouse. Cotton was the cash crop. New clothes consisted of overalls, shirts and shoes. Big bags of sugar, salt, and coffee were for winter use. The Christmas tree was cut on the farm and decorated with strung popcorn.

In 1945, I watched from a window as the power company men came to tell Daddy to hitch up the mules and help set up the poles. No more stumbling around in the dark.

Lights shined like stars all over the neighborhood. The excitement grew as time was saved from new appliances and neighbors sharing. The radio opened up communication and bonding as we gathered around it to listen to news of the war. Christmas lights were all different colors on our little farm tree.

Little did we know how much family and community life would change from that day forward by electricity.

The progress started in the house with running water, hot water, fans, refrigerators, and stoves, thus changing wash day, bathing, cooking, and cooling of food. The electric saw sped up building of homes, stores, churches, and factories.

When the power goes off today, our lives stand still.

Circle of Brightness

Big family had each other

by Lounerria Fowler
Walhalla, South Carolina

We lived on a farm about ten miles from town. There were 16 children in the family besides our parents. I was about the middle one of them. We had no running water only a well where we got our water. We would fill the old three-bushel washtub with water in the middle of the day so it would be warm to take baths in the late afternoon. We had plenty of milk and butter, chicken and eggs. We sat the milk in cold water and kept changing it so it would be cold. The 18 of us shared the outdoor toilet.

We had each other and we were happy. We were taught to read at night by an old kerosene lamp. We had a chimney for heat and we didn't have a car.

My father would drive the mule and wagon to town to get some groceries, and shoes and clothing for the one that needed it more than the rest. My mother died when I was nine, and the older sisters tended to the children.

When electricity came, it was a different state of being. We didn't have to take the old kerosene lamp from room to room. We finally got us a refrigerator, but cooked on the old wood stove a lot longer because there wasn't any electric stoves. We washed our clothes on the old wash board until we got a gas washing machine. No one could afford a T.V. because they cost too much. Business was slow because everyone farmed.

Electricity came in about 1938. Thank God for a change in the new miracle age.

Cold Pennsylvania memories

by Robert G. Greiner
North Manchester, Indiana

    I was born June 11, 1918, at home on a farm near Harrisburg, Pennsylvania. There was no running water, and the supply was from an outside pump over a stone-rimmed well. We heated water on the stove and took baths once a week in a metal tub. A two-seat outhouse served as the toilet, with Sears & Roebuck supplying the "paper."

    The house was warmed by a wood stove in the living room, which also served as the cook stove. This did not provide much heat for upstairs bedrooms. Light was provided by several oil lamps in the house, and lanterns provided light for work in the barn. Cows were milked by hand.

    In the early 1930s, my parents purchased a Delco DC generating system, with a group of storage batteries. They also did some basic wiring in the house and barn. This was a great improvement over the oil lamps.

    I walked over a mile to the one-room school through the eighth grade. I took the high school entrance test, and passed. We lived five miles from the school, and for the first two years I drove the horse and buggy to Manheim. Then Dad let me use the 1927 Model-T Ford, Four-Door Touring Car. Since very few students had cars, the girls were friendly, hoping for a ride. I went on to college in 1935.

    Then the government started the Rural Electrification Program. In due time we hooked on to the service, and the DC system became obsolete. The house and barn wiring were updated, and this was the start of many conveniences such as: electric water pumps, radios, electric stoves and heaters, installation of a modern bathroom with a toilet, etc.

    The cold outhouse became obsolete.

Circle of Brightness

Life B.E. (before electricity)

by *Alice Papendorf*
Tigerton, Wisconsin

Millions of people today flip on electrical switches or turn on water faucets and take the results for granted. And I must admit I usually am one of them. That is, until those rare occasions when my electricity goes off and I remember my early days B.E. (before electricity), and how much easier electricity has made our lives.

I grew up in rural South Central Illinois. The first 14 years of my life we didn't have electricity. I was the youngest of five girls in the family. Cleaning the chimneys for the kerosene lamps was my task — I was the only one whose hands were small enough to get inside the chimney. This was a daily task and had to be done before dusk.

> "Cleaning the chimneys for the kerosene lamps was my task — I was the only one whose hands were small enough to get inside the chimney."

Now when I'm forced to light a kerosene lamp, I wonder, "how did we ever see to study by those lights?"

I remember the wood cook stove that had to be fired up even on the hottest days in summer. Can you imagine the misery of standing over a hot fire preparing meals or canning when the temperature outside is 100 degrees in the shade? And of course there were no fans or air conditioning. The stove had a reservoir that had to be kept filled with water. This is how we got our hot water for washing dishes. The stove also heated the sad irons for ironing clothes. So cooking, canning and ironing was sweltering drudgery in the summer time. And of course keeping food safe was another problem. The ice man came only two times a week, so the ice in the ice box didn't last too long. We put milk in gallon jars and put them in a bucket, then lowered them into a well to keep it cool. My mother usually cooked just enough perishable food for one meal so there wouldn't be leftovers to keep cool.

My husband lived on a dairy farm in Wisconsin. For his family it meant not only less drudgery for household tasks, but they no longer had to milk the cows by hand. When electricity came to them, they could have electric milk machines. This enabled them to have more cows and more income. Eventually, more and more of the dairy operation was run by electricity when they obtained electric silage unloaders and barn cleaners.

What a luxury it is today to be able to turn knobs and take a hot shower or bath! And every day if you want. Back then bathing was a Saturday night ritual. Water was heated on the cook stove. Then one by one we got our turn in the wash tub.

R.E.A. was just about to come to our community in 1946, when we moved to the northwest corner of Clay County, Illinois. There we had to wait until 1948. Then we had the house wired and the poles and lines were put up.

I can still remember the thrill when the first time all the lights in the house came on.

Circle of Brightness

Toiling in the 'Dark Ages'

by Helen C. Winburn
Conway, South Carolina

Before electricity came to our home, we didn't realize we were living in the "dark ages," We were a hard-working farm family with parents we respected. I was fourth in a family of 12 children.

The boys and girls worked in the fields, tended to the wide range of animals and helped with all the household chores, plus keeping the wood-box full of stovewood for cooking meals.

Washing mountains of dirty clothes was an all-day job in the back yard. Benches were set up for the washtubs. A big iron washpot was kept boiling with lots of firewood to supply hot water for the tubs of water. Clothes had to be scrubbed on a washboard and rinsed. The water was drawn from a well. Lye soap that mom made every year was the cleaning agent. Clothes were hung on clotheslines to dry. Ironing was done by flat-irons, heated on the cook stove and switch for a hot one when one cooled down.

We had two lanterns and two lamps that burned kerosene. Lanterns were used to unload tobacco barns before daylight, feed the animals after dark and get buckets of water from the hand pump across the yard, because water for household use always ran out after dark!

Lamps were sometimes used to sit up with a sick child, or study homework. One lamp stayed in our parents' bedroom so dad could read and write in his journal and mom could tend to the baby. The other lamp was for us to do homework, read library books and prepare for bed. Sometimes the playful kids would accidentally break the lamp shade, so we studied by a smoking wick until another shade could be bought.

The Dark Ages

One of the things I enjoyed after the crops were all gathered was helping dad and the boys cut wood for the coming year. We cut tobacco-curing wood, fireplace and heater wood, and stovewood for cooking. I was a strong tomboy and could pull a crosscut saw but I loved splitting stovewood and stacking it in neat rows near the kitchen.

> **"We cut tobacco-curing wood, fireplace wood, heater wood, and stove wood for cooking."**

We had no Christmas lights, so we wrapped sweet-gum balls in silver foil, made paper chains and strung popcorn to decorate our pine tree on Christmas. It was beautiful to us.

Then a miracle happened.

Workers from the Santee Electric Cooperative began putting up electric lines in our area in 1943. Before the year was out, we had light bulbs and an electric stove for mom. We really had a celebration for the change electricity made, and so did our neighbors! Now the younger children could study by real lights, and light up the Christmas tree. Over the years, we were blessed with a refrigerator, electric irons, washing machine and other necessities.

Chapter 2

Early Initiatives

It was in this neighborly way the spirit of REA was born.

Electric alternatives limited

In the years before George Norris fathered the REA bill in 1935, small farms obtained electricity through one of three methods: luck, money or initiative.

If you were lucky, you lived close enough to town or close enough to a main power line to make it profitable for the power companies to extend service a short distance.

If you had enough money, you might negotiate with the power company to obtain electricity. Rural lines typically cost $2,000 or more per mile to serve two to five homes. As a result, power companies charged rural customers almost twice the kilowatt-hour rate as did city customers who were under no obligation to finance construction of new lines.

Stringing lines to most farms was more costly for power companies than the anticipated return on their investment. Since the ultimate choice about extending service rested with the power companies, they usually chose not to serve farmers — unless farm customers agreed to assume extra expenses.

The only other way to get electrical power to the farms came from trying to harness wind or water power by using windmills, wind chargers, water wheels and machines powered by fuel or batteries.

In 1911, the National Electric Light Association first suggested more attention be paid to agriculture and rural Americans but took no other action. In 1913, the United States Department of Agriculture published a special bulletin on farm uses, but took no other action.

After George W. Norris was elected to the Senate in 1912, he watched with interest, in the years following World War I, as farmers began taking initiatives. A few isolated pockets of farm cooperatives secured rural lines — some were established before WWI. Yet, by the time REA was formed, cooperatives still only numbered about 50.

Norris studied the success of cooperatives in Canada and determined it feasible in the U.S. He began formulating plans for government to take the lead in developing federal properties for public benefits like electricity.

In March of 1923, the Committee on the Relation of Electricity to Agriculture (CREA) was formed to promote coordination of the power industry and agriculture with the ultimate goal of remedying the lack of rural electrical service. While they conducted their studies, rural Americans continued toiling away. Some took their own initiatives with power companies and alternative solutions to electrifying their farms.

What others said about Norris:
"Stopping at a Custer county farm on a hot day many years ago with a friend, Norris noticed a young wife fanning flies away from the face of her baby. He became almost intemperate in expressing his disgust with a system that made electric fans available to most urban dwellers, but could not find ways to bring power to rural areas so children could sleep in comfort, untroubled by flies,"
— Harold Hamil, Farmland Industries

In George Norris's words:
"... frequently the stool was pulled out to the center of the room, the lamp placed upon it, and there in the circle of its light my mother plied her spinning wheel; my sisters either knit, at which they were adept, or buried themselves in their school books, and I did my lessons. It was on such occasions a closely knit family circle, earlier bound together by adversity and grief, self-supporting and undisturbed by the lack of cash and money, sensed the full happiness of home associations."

In the words of the people, here are their stories of "Early Initiatives" ....

We get 220 volts

by Richard J. Lippincott
Blair, Nebraska

In 1927, my dad and two neighbors petitioned the Nebraska Power Company to build a branch line from their high voltage line which supplied electricity to several towns between Omaha and Sioux City.

Since it took several customers per mile to make an electric line pay for itself, and since many of the farmers along the line couldn't afford such a luxury, dad and the other two petitioners agreed to make up the difference — which was $600 each.

There was one more problem. The electric lines and telephone lines had to run on opposite sides of the highway. Since there was a line of big trees along the road past our place, dad allowed the power company to bypass the one-fourth mile of trees by running the line 100 feet outside of the road right-of-way, thus saving the power company additional expense.

Contracts were finally signed in the fall, with seven other farmers signing for the service. Those signing contracts had all winter to get their buildings wired. Our buildings had already been wired for our own 32-volt plant, so we only had to make a few minor changes. My brother wired the house where he and his bride were going to live and he also got jobs wiring for neighbors.

The power line was completed by early summer of 1928. Dad made three major purchases that summer.

The first was a five-horsepower, 220-volt electric motor to run the grain elevator — replacing the old gasoline engine that was used before. The old engine was junked and the electric motor was mounted on the old engine chassis so we could move it for the other jobs if necessary.

Second, a refrigeration unit was purchased. The compressor was placed in the basement, with pipes running up to the old ice box on the back porch, where the condenser unit was mounted in the ice chamber of the old ice box. That eliminated the bi-weekly trips to town for ice.

The third purchase was a pressure water system for the house. A pressure tank was mounted in the southwest corner of the basement, with the incoming pipe running through the basement wall and running underground, out to the well, where a new pump jack, run by an electric motor was installed. The electric wiring ran through a conduit pipe, parallel to the water pipe.

We already had running water from the rainwater reservoir on the hill, so all that needed to be done was switch the cold water line from rain water to well water and we had soft hot water and cold drinking and flushing water.

What luxury!

Electricity a long time coming

by Stacy Seibert
Lexington, South Carolina

The day the lights came on in our home was a very busy but joyous moment for our family. It brings a smile to a family's face when white work trucks roll in the yard to hook you up to electricity.

To really appreciate this remarkable, widely useful gift, we have to know the history of it. We would want to give appreciation to the people who made it happen.

We would also have to imagine what life was like before electricity came to our home and community.

Greeks in 600 B.C. discovered by rubbing your fingers on a piece of amber, it would pull string and paper to you.

From England around 1600, when Elizabeth I was queen, she was served by a personal physician named William Gilbert. He discovered clothing materials had an effect when introduced to glass and wax.

In 1733, Charles Du Fay, a French scientist first learned about opposing electrical charges. Benjamin Franklin in 1752 renamed them in simpler forms as positive and negative. In 1869, Thomas Edison invented the light bulbs. We have come a long way in all the inventions these men and many more have given us. We give our deepest thanks.

Now, we must all imagine not having electricity — no T.V., no radio, no vacuum, no water. If you need to do daily

Early Initiatives

chores imagine walking 20 feet out to the well to get water every time you need to wash dishes and clothes, (no washer or dryer) bathe, mop, clean tubs, or use the bathroom (the outhouse).

If you don't have electricity, that also means no electric heater. Fireplaces need wood, how and where are you going to get wood? You would most likely have to chop it yourself and keep it stacked by the back door for easy access No one wants to fetch water and wood in the middle of the night or even first thing in the morning.

Electricity has helped us by cutting out chores we once had to do. We now have more free time to spend with our families and make more wonderful memories.

The day the lights came on in our house was a dream come true. The MCEC man stepped out of his truck and got to work. He fixed our meter and so on.

It was the sweetest moment, knowing our home will be lit up very soon.

Power company politics

by Helena Kaczynski
Tucson, Arizona

I was a child in rural Michigan during the "Dark Ages" (before electricity). In the early 1930s, a large electrical company ran a power line toward the main road where we lived but since we were a half-mile from our nearest neighbors, the line stopped short of our farm. To continue the line, the power company required three houses or farms per mile.

One "skin-flint" neighbor had his farm wired and invited his less fortunate neighbors to see his light. He said that if we wanted cake and coffee afterwards, we should bring the cake (another neighbor brought coffee.) In his kitchen, the host picked a bulb from the table, climbed a ladder, and screwed the bulb into the ceiling lamp. He turned on the light. What wonders to behold! Then he reversed the process and laid the bulb on the table. We went into the living room where he repeated the process of ladder, bulb and light switch.

My dad asked, "Is that necessary?"

"You can't tell me that turning off the switch on the wall will stop the electricity from flowing out of the light on the ceiling. I'm not wasting electricity."

Years later, the REA ran power lines in our area. The large electrical company, which 15 years earlier had refused to run their line to our farm, asked dad to stall the REA by refusing right-of-way. That way the large company could beat the REA to most of the farms in the area. Dad refused.

"Why should I do so when I begged you and you refused to run your line that extra half mile to my farm 15 years ago?"

And so the farms in Montcalm County came out of the dark, not just on the main road, and not just three houses per mile, but every farm, thanks to the REA.

REA: Government success

by Bert Stanke
Phillips, Wisconsin

It was a long, hard struggle to make inventions like Edison's light bulb usable all over the United States. In the beginning, utilities were eager to wire the cities, but rural areas were a no no.

My mom used a gas engine washing machine and kerosene lamps while the family milked cows by hand on our small hobby farm. Dad's main income was teaching grade school. We pumped water by hand for us and the cows.

My dad circulated petitions and tried to negotiate with utilities for electric line. He tried getting farmers to agree to donate right-of-way land, clear it, and set the poles if utilities would agree to string the wires. Their answer: not for less than four customers per mile. This went on until President Franklin Roosevelt with the help of legislators like George Norris of Nebraska passed the REA Act.

It still took time for men like my dad to help organize cooperatives, but they did. And eventually all of America was electrified and we had electricity in our homes.

But I think of how hard they fought for us and that makes me a strong supporter of cooperatives and credit unions When I hear some pseudo-conservative fool ranting and raving about failed government policies, I'd just like to get them in my car and show them as I drive past miles and miles of electric lines — built by REA. And miles and miles of fire lanes and public forest built by Civilian Conservation Corps and dams and parks built by Work Projects Administration which our tourism and wildlife depend on.

And tell them if that's government failure, we need more of such failure.

Circle of Brightness

Farmers formed own line

by Alberta Wolf
McCook, Nebraska

My thoughts go back to the early 1910s where I was born and raised in Bartley, Nebraska.

Imagine if you can, a dining room with a table to accommodate six children ages three to 17 and their parents. At the beginning of twilight there was homework from school, reading or mending to do. A glass kerosene lamp sat in the middle of the table to afford light for these activities. Many hours were spent filling lamps with kerosene, washing the smoked glass chimney, and polishing them crystal clear for the rays of light better to reach the written words.

At bedtime there were smaller lamps for each bed room. Much care in situating these lamps at bed time were a concern for our parents. Whether or not we younger ones were lighting our lamps, we were cautioned to "be careful." When the two older children were small, mother lit a match in a dark walk-in closet to find an article. They determined later that the head of the match flicked off with a spark still in it. Hours later it ignited fabric or a scrap of paper and soon the clothing burned. Luckily the fire was contained. A lesson learned was forever with us; the "be careful" warnings came from the heart.

Life without electricity meant making plans to the best of our ability for the more important things to be done in daylight. Emergencies were very difficult in darkness and by the dim light from a kerosene lantern. There was still danger of fire. Time — in the absence of electricity — was of the essence. Much time was spent in upkeep of lamps and lanterns.

In the late 1920s and early 1930s came a revolution in the event of electrical power. A group of farmers southwest of Alma, Nebraska, in Harlan County, formed a "farmers line" to bring electricity to the few farmers south of the Republican River. Volunteer workers were called for. I think my husband Leo Wolf was the first to volunteer and he worked feverishly so that it might be finished by the date of our wedding.

It was done in time.

So along with the new role of housewife, I was introduced to the use of electricity. After living with kerosene lamps it was a wonderful life. There were lights all over the house. We even had a clothes iron and an old electric vacuum cleaner. Who could ask for more — or in our wildest dreams who could expect anything more?

From that day forward, we praised George Norris for his part in REA when it replaced our "farmers line." Many years later, we visited the area around the Tennessee River Valley and Norris Dam where the concept of REA was born. My husband Leo took great pleasure telling tourists and even local people about "Norris the man." Some local people did not realize the importance of this man.

I will always respect rural electricity and the freedom for which it stands.

> **"I think my husband was the first to volunteer and worked feverishly so that it might be finished by the date of our wedding."**

Circle of Brightness

Pre-REA days simple

by Mary Watters
Tecumseh, Michigan
(As told by her father, the Rev. Donald R. Baughey)

Back in 1916, growing up in a big family of six boys and two girls, I well remember the days of no electricity. We had oil lamps in every room of the house. How vivid I remember the big fancy one on our living room table. It was the centerpiece of the room.

One special lamp had a bag-type hanger on it. We had to pump it up and when it was lit, it gave off a bright illuminated glow from the reflection behind the blade. We always had to keep the globe clean and the wick trimmed so it didn't release a lot of black smoke into the air.

I can also remember going to bed knowing very well that before morning the oil would be gone, but no one wanted to get up and refill the lamps. So we would all try to go to sleep before they would burn out. I was so afraid I always wanted to hold my big brother's hand. I would call to him many times so I could hear his voice and make sure he was still awake. If he answered me, I knew I was okay. If he went to sleep first, I would be in trouble.

> **"I was so afraid I always wanted to hold my big brother's hand."**

In the mornings it was always dark too as we always got up before daylight to a pancake and sausage or eggs breakfast. I remember mother standing over the old cook stove making plate after plate of pancakes for her boys. No General Electric® stove in her kitchen. Not even a microwave oven. This was followed by prayer and Bible reading, a big part of our daily life.

I also remember mother sweeping the rugs with no electric sweeper. She would sprinkle the rugs with water to keep the dust down and then she would sweep and sweep the rugs. We had never heard of the power Dirt Devil® if you please.

Our evening chores included making sure the kindling box was full and the coal pail filled. This had to be done before the sun set, because it was no fun after dark. We did not have a mercury yard light to help us.

For our entertainment, we never played T.V. games or even knew what a T.V. was. One man strung a wire in the tree to use as an antenna so we could hear the radio. If we were real quiet we could hear the announcer talking. It was very hard to keep six children quiet.

Another one of our fun things we did was to shake the popcorn pan over the old cook stove and hear it pop, pop, pop! Add salt and butter and chow!

I'm glad I lived through the good old days with all the memories of everyone helping each other. However, I'm also glad for today and the power of electricity.

The day the excuses ended

by Harris Shuman
Canadys, South Carolina

We lived in a small community with the Southern Railroad 200 feet from the house. Highway 78 ran 150 feet parallel from the tracks. At a young age, light lines were put along the highway. Families on both sides were getting lights for the first time. Dad tried about four years to get electricity but they kept saying it was too much trouble to cross the highway and the railroad at the same time. They claimed you had to have extra long poles to cross a railroad. Railroad men said they did not want power lines across their train tracks.

Dad heard if you bought a refrigerator or electrical-powered machinery, they would have to put in power for you. Everyone in the family saved money to help buy a refrigerator. We put it in the kitchen with no electricity to run it.

Then the excuses really began.

Finally when we came home from school, they were putting the lines across the railroad. We had a big family and everyone was thrilled and more than willing to help put wiring in the house. We had a porch on the back about half-way down one side. Only one light was installed where the porches met. One light was put in the center of each room. All were with the pull cord on them.

Finally the day came when the lights would be turned on. As night came, everyone went from room to room pulling the cords. We were all amazed what light did to our house. Us kids tried to help Dad pay the light bill. We were not going to give up those lights. This was the middle 1930s.

All my life I have enjoyed the many conveniences of electricity. But most of all, I still don't want to give up the lights.

We provided poles, wire

by Harvey B. Burkholder
Grayling, Michigan

In the Fall of 1934, we moved back from the city to the Burkholder homestead farm four miles north of Mancelona, Michigan. After living a year without electricity, we had a desire to have it on the farm.

The main electrical power line provided by the Consumers Power Company ran about 1,500 feet from the house. They would not run a line to the house without a considerable charge. Money was not available. The electric company proposed that if we could provide the poles and wire they would string the wire and provide the electricity.

This seemed an almost impossible task. However, after much deliberation, we came up with a plan. Brother Milo had been working for an electrician in the area installing electricity in homes. Therefore, he had the necessary experience for the installation of wire, boxes and light fixtures required in our farm house.

Growing up on the farm, along the banks of the Cedar River, were large cedar trees. We cut a few down, stripped the bark, and horse-hauled them to locations along the road leading to the main electric line. Holes were dug by hand-shovel and the poles placed in the holes and were set.

The heavy copper wire needed for this main line was purchased from brother Milo's meager earnings. The wire was strung on the poles, the house was wired, the electric company made their connection. We had electricity in the house.

Now, mother could use her washing machine, brought along from the south. We could use the radio, toast the bread and most importantly — enjoy better lighting throughout the house.

Circle of Brightness

City extended electric lines

by Lester C. Mohling
Glenvil, Nebraska

It's 1928 and I am six years old.

We were the first rural residences outside the city of Hastings, Nebraska to have an electric line built to our home. The city would only build a line if there were three customers per mile. The first line was only two and a half miles long. Our house was in the first mile. One resident refused to use the service because they owned and used a Delco® electric plant. Later they did hook up to the Hastings line.

"I'm just too small to reach the light switch."

That is the excuse I gave because I could not turn the light on.

> "I'm just too small to reach the light switch."

When our farm received electric service each room in the house had one light cord hanging from the ceiling with a turn button switch — so I needed to stand on a chair to turn on the light. Only one outlet receptacle was installed, for the radio which was the family's main interest. The line to the barn was made with a single light in each horse stall, two lights in the big milk room, and a light in the hay mow. Before electricity, we used kerosene lanterns in the barns. Gas lanterns in the house hung from hooks in the ceiling.

My mother was so delighted with the electricity because she was given an electric motor for her washing machine and an electric flat iron. Of course more appliances and advantages were added later, such as a line to the milk tank house and a line to the brooder house, for the automatic electric chick brooder stove.

Our school house was not to get electric service because of only nine months desired service. Some four or more years later the main line was extended and the school house did get electricity. There were two lights hung in the big room and a light put above the front steps, all with wall switches.

There was not much more than lights and small appliances in the rural homes for several years. Then at the time of World War II, there were lots more electric services realized by rural families.

Now in these same areas all things are electrified, such as irrigation and home water wells, street and yard lights, farm shops and heaters for stock water tanks and small baby animal brooders.

The change from years gone by is phenomenal.

We are all thankful for this thing called electricity.

Circle of Brightness

Company wouldn't build line

by Naomi Wingard Hite
Gilbert, South Carolina

I was born about eight miles south of Gilbert, South Carolina in 1920. My parents were not wealthy.

As children, we walked to our little community school with two teachers. The girls swept the floors at school and the boys went in the woods nearby to cut firewood for heating. We had an outhouse as a toilet. We brought water in for drinking.

There was plenty of work to do at home. We had to bring in firewood for heating and cooking. We studied our lessons by kerosene lamplight. Our home was drafty and had no insulation.

I have seen our drinking water freeze while setting on a shelf.

I was third from the baby in our family, so I didn't usually have to get up and make the fire in the fireplace on those cold winter mornings.

We had chickens to feed and a cow to milk. We would butcher a hog in the winter when the weather was cold enough to save the meat, which would hang in the smoke house to cure.

When I was in high school, my father bought a kerosene-burning refrigerator that would make ice. We would light the wick each day and it would use about a quart of kerosene each evening so we would have ice the next day.

At that time, I thought we were really coming up in the world.

When I got married, my husband and I lived at the intersection of Juniper Springs and Two Notch Roads, only a mile from Gilbert. A private company supplied the town of Gilbert with electricity. They would not build a line to serve

Early Initiatives

our area with electricity. My one and only daughter was born on a cold January night by lamplight.

Some time later we heard about the Federal government passing the Rural Electrification Act.

That was the source of funds for the cooperatives to finance rural electrification. When Mid-Carolina came in to being, we signed up to get the lights for a minimum of $3.50 a month. We were really excited when the Co-op lights were turned on. That was just the beginning of what was to come.

First we purchased an electric refrigerator and gradually added more appliances, such as an electric range, electric iron, washer, and an electric water pump and hot water heater. It took a while to accomplish all of this, complete with indoor plumbing.

We had a lot to live for back then, saying nothing about the new technology of today. It has changed our way of living entirely.

> **"They would not build a line to serve our area with electricity. My one and only daughter was born on a cold January night by lamplight."**

Chapter 3

The Rural Countryside

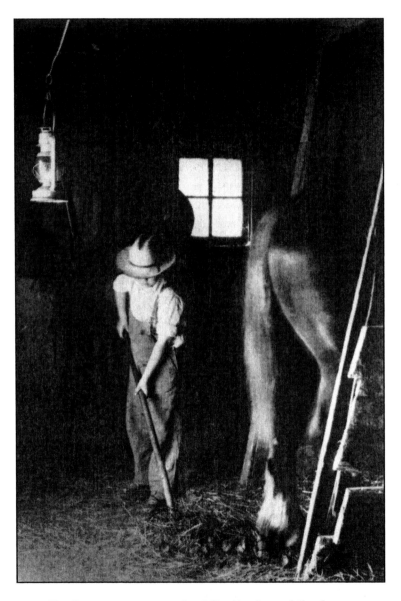

Darkness was a constant limitation of the farm.

Circle of Brightness

An appreciation of farm life

After George Norris's father died and his mother was left to run the farm, young George performed the usual chores expected of a farm boy and more.

He had long, strong hands and as soon as he was capable, he worked by day in the harvest fields of neighbors. He usually charged a dollar and a half per day and board, but once, a reluctant farmer offered George two and a half cents a bushel to harvest a field damaged by wind. George took the job and worked so hard the farmer wound up paying him $2.55 a day.

Later when Norris became a district judge in Nebraska's homesteading days, he presided over a large number of liens and claims against farmers whose crops had failed. The empathetic farmer-turned-judge developed a common-sense approach to these cases brought before him. He determined that if a debt-ridden farmer could, under ordinary farming circumstances, meet his indebtedness, he would postpone confirmation of the sheriff's sale and give the farmer an opportunity to pay it.

When Norris first ran for a seat in the House of Representatives, he was opposed by a man who claimed to be a farmer, but Norris believed his main interest was in running a bank. Likewise, his opponent levied charges that Norris was not a farmer, but an attorney. At a political debate in McCook, Norris challenged his opponent to a cornhusking contest to demonstrate once and for all who the real farmer candidate was. Norris proposed the two candidates go out into the field the next day and husk corn from sunrise to sunset, stopping only for an hour dinner break. He said if he could not husk more corn than his opponent he would withdraw his candidacy if his opponent would agree to the same. His opponent did not.

The Rural Countryside

Shortly after he was elected to the Senate, George W. Norris spoke before the House Committee on Agriculture about his resolution to provide for the appointment of a Farmers' National Cooperative Commission.

With his farming background, he was always looking for ways to make rural living more attractive and believed this was a step in halting the number of farmers leaving the countryside in favor of more prosperous opportunities in the city.

He continued looking for ways to make electricity available to rural Americans.

What Others Said About George Norris:
A Democratic newspaper, The McCook Times, endorsed Norris for his first Senate campaign saying, "He was born and raised on a farm, and all his interests are now and always have been connected with the farmer and those who toil."

In George's Words: "From the time I grew old enough and strong enough to work, that farm fully occupied my thoughts and my energies. Its acres were the most exacting of taskmasters, seeming with a chore for every free hour of the daylight. From daybreak to nightfall the farm cried loudly for attention completely absorbing the strength and the thoughts of a growing boy."

In the words of the people, here are their stories of life on the "Rural Countryside"....

Circle of Brightness

The non-electrical farm era

by Max C. Scheller
Ogema, Wisconsin

1900-1941.

Anything we did in that period of time was done by daylight, lamp or lantern — maybe some candlelight. Such as the care of animals, barn work and milking, feeding, and preparing the horses to do the woods work which was some income in addition to a small heard of cows. The kerosene lantern was the only source of light to do these chores. Milking of course was hand milking. One would take a pitch fork and lantern and go to the mow to pitch down hay for the cows and horses.

We cleaned barns by a wheel barrow wheeled on a plank to the manure pile — later a bucket on a cable and out to the pile. A third more up-to-date method was a bucket on a track with a 40-foot boom to swing it where you wanted the pile and anchored in a specific area.

Putting up hay in the 1920s, we had a carriage on a track in the peak of the barn, with a fork and ropes to take the hay up into the mow. This was pulled by a team of horses. After rural electrification, a motor-driven hay hoist performed this task.

Most families at that time were large, 10, 15, even 17 children. Often, older children were gone from home when the younger ones came along. Food preparation was

> **"Putting up hay in the 1920s, we had a carriage on a track in the peak of the barn with a fork and ropes to take the hay up into the mow."**

The Rural Countryside

done on wood fired ranges which had large ovens to bake many loaves of bread. Reservoirs on stoves held six gallons of always hot water. Bathing was done by the range in a galvanized wash tub with water heated on the stove. We had a wash stand with wash basin, soap and towels for hand and face washing, and a pail for waste water underneath, emptied outdoors when full. Clothes washing was two wash tubs, a washboard for scrubbing clothes and a handwringer. One tub was used for washing, the other for rinse water. The white clothes were boiled to keep them white. This was done on a copper boiler on the wood stove.

    The well was hand dug — 80 feet deep. A hand pump later or pump jack and small gas engine to run it. This provided water for the house and barn.

    We also had a welder with a generator run by a tractor.

    Most things were home made and tools were made in our own blacksmith shop.

    We had a farm flock of 30 chickens, and a rooster or two to fertilize for hatching eggs.

    In cold weather, two lanterns were hung to warm the building. Today, this is warmed by a heat lamp.

    After milkings, milk was taken to the house where it was separated by a hand-poured cream separator into cream and skim milk. Cream was sold to a creamery or made into homemade butter. Skim milk was fed to calves and hogs.

    There was the outdoor toilet, which was used in all weather.

    What a pleasure to receive rural electrification in July of 1941.

Circle of Brightness

Bride went back in time

by Lillian Barnes
Cotesfield, Nebraska

Having been born December 14, 1908 and living under 17 presidents, oh what drastic changes I've experienced during these years.

In 1928, I was a young new bride coming from town to live in a rural area. I was used to modern conveniences.

Instead, I found myself filling lamps with kerosene, washing and shining lamp chimneys, swinging the handle of a washing machine filled with water heated on the wood-burning stove to clean clothes, and using sad irons heated on the wood burning range.

I found myself churning cream into butter using the stomping method. A long wooden handle that had a paddle attached to one end fit into a barrel-shaped crock, and it often took 20 to 90 minutes before the cream turned into butter.

> "I found myself churning cream into butter using the stomping method."

I found myself cranking the milk through the separator by hand, pumping water by hand for everything water is used for, carrying buckets of water to my thirsty flower bed, making ice cream in a hand cranked freezer, carrying a lighted lamp from room-to-room, quilting by lamp light and using a cold outhouse in the winter time.

How thrilled we were the day the workmen pulled into our yard with their trucks loaded with wire, long poles and equipment to begin the wonders of electricity. We did the evening chores earlier than usual, and when it was dark we turned on every light in the house. Then we went outside to view the wonders from the outside, as well as the inside of the house.

My husband, upon hearing that cows gave more milk with music, took my little kitchen radio to the barn, stationed it on a shelf, and played music while we were milking. I can't honestly remember if we got more milk than our usual ten-gallon cream can and two buckets of milk. How nice to have lights in the barn instead of a lighted kerosene lantern hanging on a peg.

When I was a child, before electricity, the lighting in our home was carbon gas lights which did not put out a very bright light. When my brother needed an appendectomy late one evening, I watched my father hold a kerosene lamp close while the doctor performed the appendectomy.

No, I do not wish to go back to the "good old days."

A special big thank you to the memory of George W. Norris, the United States Senator from Nebraska, the father of rural electrification, which contributes to my ease and long life.

Circle of Brightness

The 135-pound man

by Ron Moser
Missoula, Montana

Before we got REA, we had an antiquated 32-volt light plant that ran on 24 batteries. It was supposed to be easily charged with a 30-foot windcharger beside the garage.

While the wind never seemed to cease on our Brady, Montana farm, the windcharger couldn't handle the challenge. So we also ran an auxiliary motor stationed right beside the house. It was one of my jobs to keep the motor gassed to run all day so we could have enough juice to run a converted ice box and keep a single overhead kitchen bulb burning in the evening.

The only thing the windcharger seemed to be good for was to teach me to crawl up and learn to swing by my hands — which gave my mother substantial worry. The constant noise of the motor, not to mention the disgusting exhaust polluting the air outside the back door, nullified the comfort the night light gave. Yet the importance of keeping food fresh ensured the survival of the light plant.

Another one of my jobs was to take my wagon to the barn, the center of which was an enclosed 12-by-12 room filled with sawdust covering cakes of ice. I would use ice tongs to slide a cake of ice down a 2-by-12 onto the wagon, then pull it down to the house where my dad would carry it in and put on top of the ice box. During the day, the ice would melt and drain into a pan on the bottom of the ice box. It would also be my job to carry that out in buckets for the chickens, because not only did we not have electricity, but we had neither running water, nor a well.

We hauled water, the closest source being 12 miles away on the Teton River. My Dad's 1937 Chevy pickup truck struggled with the 300-gallon tank, so we really conserved water, down to the proverbially sharing of the metal tub on the kitchen floor for the Saturday night bath. Also, because of the limits of the 1937 Chevy, we would pull a trailer made from a Model-T frame back to the Brady Irrigation ditch a mile away, filling 55-gallon barrels to supplement chicken and stock water.

During the winter, my job was to pull water from the cistern in buckets to carry to the chickens and to the house. I also had to chop and carry wood and coal into the house for cooking and kitchen heat, until the magical day my dad found an oil-burning range. That was a monumental change.

About the same time we got electricity, Brady's town fathers raised the money necessary to put in a water and sewer system. The war had been over long enough so we could finally get new big trucks. Dad bought a 1947 Studebaker 1 1/2-ton and a 1,000-gallon tank, and I had a new job of hauling water only three miles away in town. I could pick up another $20 spending money every couple of weeks from each bar hauling their beer cans and bottles to the dump using the truck hoist.

That hoist also made it easy to later haul 6,000 bushels of wheat three miles to town in one day, along with a second truck. In the olden days, Dad was very proud of hauling 900 bushels of wheat to town in nine hours on a Model A, one-ton truck, all of which he had to shovel from the floor. That's 10 pounds a minute he averaged, shoveling for nine hours straight!

I'm damned proud of my 135-pound dad for that, too, and I think that alone illustrates what electricity did for the common man in rural America.

Circle of Brightness

Food storage was chore too

by Bea Wester
Catawba, Wisconsin

Kerosene or gas lanterns were used in the homes and barns. We were very leery about using them — especially in the barns. Light from them was quite dim and difficult for reading.

We washed clothes by hand in water carried from the well and in earlier times, from the river. Later, a windmill pumped water to the barns. On calm days, we pumped by hand for about 12-15 head of cattle. Water was stored in a large tank above the ceiling of the cow barn for the animals.

Dad rigged up a wind charger on the roof of the house with one wire running down to the kitchen. It supplied power to one light bulb. A strong wind produced a bright light — a weak one produced a dim light. This system included a car battery for storing power, but it didn't last many hours.

We stored fresh meats and fish in a 25-foot deep hole in the ground. These were lowered down in pails. Butter, milk and eggs were stored on the cool basement floor. Milk was stored in glass jars surrounded by cold water, which had to be changed often. This meant many trips down there between meals.

> **"We stored meats and fish in a 25-foot deep hole in the ground."**

Later, we used ice boxes in which large blocks of ice were stored in the top to keep things cool. Smoked hams were wrapped in cotton dish towels and paper to keep flies, etc. from getting to them. These were pulled high up to the hayfork track in the hay barn and stored there until about August 1. None ever spoiled for us.

The Rural Countryside

Things changed in the late 1930s when we first got electric lights, then washing machines, milking machines, refrigerators and radios. Water was pumped by gas engines or electricity. We had no running water to the house until the late 1950s. Bathrooms weren't installed until much later.

You see, barn conveniences came first. Things for the house came later.

Circle of Brightness

Lantern didn't light dark barn

by Janet Brunswick
Bryant, Indiana

I remember when my dad would light the lantern every morning before he would go to the barn to milk the cows. Our old barn was as black as the "Ace of Spades" and he needed some light to see the cows. He told me it would be nice to have electricity in the barn.

The year was 1945. I was five and wanting to go to school, but I was at home every day while my brothers and sisters were going to school.

One day a black truck pulled in the yard and a man took things out of his truck. He headed straight to the barn. Of course, I was interested so I went to see what was going on.

That man was nailing wires to the beams of the barn and installing little boxes. Dad told me we were getting electricity installed.

That didn't mean much to me so I went to the house and played. I forgot all about what was going on in the barn.

> **"Not only did we have light in that old barn, but I never forgot that smile on my dad's face!"**

That night my Dad took me by my hand and said, "Let's go to the barn."

When we got inside the door he switched on a button and I stood there in awe. Not only did we have light in that old barn, but I never forgot that smile on my dad's face!

Our house had only one light switch and one light bulb in every room. We thought that was like heaven.

Hard work powered dairy

by Julian M. Johnson
Mason, Wisconsin

We milked 25 cows by hand, plus feeding chickens, pigs and horses. We did chores by kerosene lantern. Barns were cleaned with a shovel and the silage was taken from the silo by fork. Water was pumped in a large tank by the windmill. When there was no wind, a gas engine was used — if it would start. If not, we used a farm tractor or pickup truck with the belt running off the rear wheel. We cooled milk in 10-gallon cans by the well. There was no running water.

The house was heated by wood and the cooking done on a wood stove by kerosene lamps. The washing machine had a temperamental gas engine. Our shop had a blacksmith forge, a hand-operated post drill and other hand tools.

The year was 1944. I was 16. I had to quit school two years earlier. We were promised electricity in 1941, but war changed that. My only brother was on a free camping trip across Europe with General Patton, so Dad and I ran the farm.

The shop was the first to change with electric drills, grinders and table saws. Milking machines, bulk tanks, silo unloaders and barn cleaners followed later on. Running water was added to house and barn, and indoor plumbing in the house. Later on came central heating, air conditioning, electric stove, TV, microwave ovens and refrigerators and freezers.

Electricity was the biggest change for rural America. Every year we added something new electric, making life easier and allowing us to produce a better product, but the best of all was an electric welder and battery charger.

We thank George W. Norris and other people that pushed rural electricity, for many senators of that time said the kerosene lantern was good enough for the rural people.

Circle of Brightness

Ranch life for farm wife

by Aldine Booco Fiedler
Grand Junction, Colorado

I was born in 1930 at Copper Spur, Colorado, which is no longer a town. My grade school years were spent on a ranch between McCoy and Toponas. We used kerosene lamps or gasoline lanterns, cooked and heated with wood or coal.

There was no electricity near McCoy, Copper Spur, Bond, State Bridge or the Sheep Horn area. My oldest son David was born in 1950. We used carbide for our house and barn lights. When the lights went out, they put a can of carbide granules in the hopper, and the graduals dropped gradually into the water tank below, forming a gas that flowed through the tubing that was in every room of the house. One light to a room. It was not a bad light, but in the middle of a company dinner, the light may go out.

Washing clothes, sheets and blankets was a story in itself. Mother washed on a washboard until 1941. We got a washer from Montgomery Ward® that had a Briggs-Stratton® gasoline engine. That is what I used on the ranch. It was so hard to get the engine going that I was tired by the time the washer started. Ironing was done with irons heated on the wood stove.

We attended many meetings, talking with REA officials, trying to get service to our area. The REA lines started from Toponas about 1950 and marched along through McCoy, Copper Spur and Bond, then up the Colorado River over the rough terrain to State Bridge, Sheep Horn and Radium.

Daughter Jackie was born in 1954, still no electricity, but the line was on the way. We watched the progress with anticipation. Jackie was three months old and my house was a mess with electricians running wires through the 14-room log

The Rural Countryside

house getting ready for the big day, which occurred in the summer of 1954.

Then the lights came on.

I cooked three meals a day for ranch help and the family. When cooking on my new electric range I would stop and think there was something I should do before continuing with the dinner; it was putting more wood in the stove, or waiting for it to cool, so the heat was right.

> **"I cooked three meals a day for ranch help."**

A lot of things the REA made possible for us. The 200-foot sheep shed, where the lambing ewes were kept with their lambs, had light. The barn where the milking was done had light. We could hardly believe how much easier and more convenient life could be. The furnace for this large home was converted to a stoker run by an electric motor. We had a warm house all the time.

Now we live on Orchard Mesa, Grand Junction. We have electricity furnished by Grand Valley Rural Power. Most people these days do not even know about going without electricity. It is still a source of enjoyment, and I am thankful for the comfort and ease it brings to my life.

Circle of Brightness

Maine city kids try the farm

by L. Vincent Baud
Scottville, Michigan

In early 1940, rural Maine — like parts of rural Michigan — was without electrical power. My sister and I were city kids growing up with all the benefits of electric heat, lights, and labor-saving appliances; nevertheless, when we took the train north to our grandparents' dairy farm for summer vacations it was a leap back to a time before electricity.

Utility poles never made it to the farms at the end of the road. Sis and I were allowed to lollygag until 8 o'clock most mornings. Uncle George, however, was up at four loading milk cans into the truck for a daily rendezvous with the train. Grandpa Joe and Uncle Bill were out in the barn hand milking 17 cows by the light of kerosene lamps. Meanwhile Grammy had a hearty country breakfast cooking while water for washing heated in the stove's three-gallon tank.

Life on the farm in those pre-electric days was always a challenge. Our grandmother washed by hand every Monday, scrubbing the clothes with homemade lye soap on a corrugated washboard in a large galvanized tub. This same tub served very well for the ritual Saturday night bath — whether it was needed or not. I helped Gram by turning the handle while she fed the laundry through the wringer — twice. The clothes were strung on lines outside in good drying weather, inside the wood shed on damp days. Sheets and shirts were spread on the grass for the sun to bleach. Ironing was done on Tuesdays. I can hear the clank of the handle as she changed irons on the hot stove.

Evenings brought us into the kitchen, reading and doing picture puzzles by the light of kerosene lamps. War news and music filtered through A.M. static over a battery-powered radio. By 9:30, all were ready to retire because of fatigue and eyestrain due to poor light. We climbed the stairs to our respective rooms and by lamplight our shadows climbed with us, flickering on ceilings and walls. All too soon we bid our sad farewell to summer. For us, it was back to school and a city life of electrical conveniences. For them, another phase of living was about to begin.

> **"For us, it was back to school and a city life of electrical conveniences. For them, another phase of living was about to begin."**

Fall was a time to sharpen saws and hone axes — readying the woodshed for next year's supply of birch, maple and oak. That shed was conveniently located between the privy and kitchen and you were expected to gather an armful of wood on your way back. When the pond froze, ice would be cut and hauled to the icehouse in 400-pound blocks. Fifty-pound chunks were cut and dunked in a water tank out in the milk house where the five-gallon cans cooled.

One day my Uncle dumped the milk because of a blizzard. He called it quits and sold the farm.

The very next year power lines came through replacing the milk house, well house, and ice house, all with that one-labor-saving device we call the refrigerator.

Farm kid chores unending

by Donna Akeson
Chappell, Nebraska

"Hey kid, go fill the water bucket"

I remember it like yesterday. I lifted the white enamel bucket rimmed in red, removed the drinking dipper and started for the windmill counting my steps . . . one, two, three . . . 81!

Once there I lifted the lid from the wooden pickle barrel centered in a dirt encasement, gently dipped my bucket in, and heaved it up. I had to be careful not to disturb the string off to the side that was attached to a fruit jar of milk which sat on the bottom of the barrel. Loosing that string could mean a chilly dip to the bottom at supper time so we could sip cool milk as we ate around our table that was dimly lit by a flickering kerosene lantern.

As time passed, Dad wired the house and installed a generator in the old granary out back. I was pretty excited, that was before I learned about . . . THE BUTTON.

I found it ironic that having a modern convenience meant one more chore for me. When evening fell, we would flip on a switch that was supposed to start the generator which rarely worked.

> **"I found it ironic that having a modern convenience meant one more chore for me."**

"Hey kid, go to the granary and push the button."

I braved the darkness complaining to myself but careful not to make a sound to alert all the lions, tigers, bears and boogie men waiting to nab me. Quickly I pushed . . . THE BUTTON . . . turned toward home with my feet flying, leaving

The Rural Countryside

all the evils that pursued me in the dust.

"Go push the Button," were the most feared words but I was comforted that we could do dishes with a bright light and water heated on the kerosene cook stove.

It wasn't long until I began hearing comments like, "When the lines are in we will get a pump jack and have running water," and "when the lines are in we will get an electric range."

I wasn't sure what they meant, but my heart was sure we would be living like movie stars. Dad installed a hot water heater in one of the closets and started adding a bathroom with a big shiny white tub to replace the old tin one and my favorite improvement, a commode. Many long weeks they sat there taunting us as if to say, "when the lines come in. . ."

Soon tall poles were set in, lining our road like soldiers at attention with wires draped from pole to pole.

The day they hooked us on was magical. I can't describe it. Not only did the lights come on with a flip of a switch, the pump jack pumped water through pipes into the house, the bathtub filled with a turn of a knob and best of all the toilet flushed. I know I was happiest of all... because I felt just like a movie star and the water bucket I had used to carry water in and . . . THE BUTTON . . . were now a part of history.

Electricity changed life

by Howard P. Baker
Orleans, Indiana

Where was I when the lights came on?

The switch for me was turned on November 2, 1939. It was really a big change for me. At the time, just my mother, Dora Baker, and I were living on the farm. My father died in 1936. We turned the lights on for the first time about 4 o'clock in the afternoon. What a change to be able to turn on a switch instead of having to light the old kerosene lamps.

This farm has been in the Baker family since 1820 when it was purchased by my great, great grandfather, Frederick Baker. The property is in the Northeast Township of Orange County Indiana. Currently it is owned by our daughter, Clora Mae Baker, and at present the farm includes 120 acres.

Rural electrification was the greatest invention ever for the rural country. Two men wired our house, and my mother and I were so proud of their work. We felt like we were really modern — just like the people in town.

> **"Rural electrification was the greatest invention ever for the rural country."**

Rural electrification changed our lives in every way imaginable. After the electricity was turned on, we began to change to modern equipment — the electric range replaced the old wood stove. That meant that I didn't have to chop and carry wood for cooking any more. We bought a wringer washer and eventually a refrigerator. What a luxury.

After I married Bethel Newlin in 1942, we began increasing our Jersey dairy herd and eventually bought our milking machine and milk cooler. We had Permit No. 1 for

Grade A Dairy in the county. We were in the dairy business for 40 years. We finally decided to sell the herd and retire when I was 75 — that was 15 years ago. I have lived on this farm my entire life and plan to continue living here for as long as the Good Lord lets me.

As we look back over all of the inventions and changes that have occurred in our lifetime, without a doubt the most significant was rural electrification. I will never forget that day in 1939!

Our farm way of life changed

by Leslie Hohndorf
Shelby, Nebraska

How I remember July 8, 1938!

On that evening I was in the field south of our house cultivating corn with a team of two horses and two mules. I remember looking up toward the house. The house was lit up more than usual. We used kerosene lamps and a gas lamp when somebody called on us. Naturally, I thought we had company for the evening. I remember studying the building sights to try and see who was visiting that evening.

All of a sudden, the yard light came on!

Then I knew — our long awaited electricity had been put into action. We had our place wired for electricity in the spring of that year. For this 15-year-old, there was no more cultivating that evening. As soon as I got to the end of the field, I headed Queen and Pet, the horses, and Bill and Jim, the mules for the homestead. Everyone was so excited!

We sold our gas lamp for $1 that week — not everyone got electricity right away.

How things have changed. From kerosene lamps, a gas lamp, candles, and one battery-operated radio, to a whole new way of life. Before we had electricity, we cooked on a wood burning stove even when it was so hot out.

We also heated our water on the stove along with the reservoir on the stove. A bath came once a week - on Saturday night in the kitchen, with the heated water from the old time cook stove. It took time, but changes did come eventually.

Now look at the changes and progress. My generation has enjoyed a lot of progress and change in our lifetime.

A double-barrel shotgun shack

by Levi H. Fox
Salley, South Carolina

    We lived in Aiken County, I was 13. Daddy worked a two-mule sharecrop farm about 75 acres. We worked from sunrise to sunset, the best time of my life. We lived in a four-room, double-barrel shotgun shack.

    Our light then was an oil lamp that you carried from room to room. Daddy got a man to come and mount a meter base, a 60-amp. panel and wired four pull-chain lights in the ceiling one in each room.

    Then we waited two or three months for the power company to bring power to our house. Then one day the lights came on. I remember it well. It was 1953 when REA was busy building a new way of life for the country people.

    I can still hear daddy say. "Boy didn't I tell you when you leave a room to turn that light out." Back then we read our own meter and sent it in.

    Our light bill back then was about $3 a month. A real bargain for light back then.

Circle of Brightness

Electricity dissolves isolation

by Beth Ayres
Crawford, Nebraska

Our family lived in the middle of Cherry County in Nebraska's Sandhills. My father, Paul Weber, was born and lived all his life on that place 50 miles from Valentine and 40 miles from Thedford — our closest town. Mostly, we didn't go to any town; we just went to school and back. Usually, we did so by walking or riding horseback. We all worked hard and found plenty to do.

Our roads were county trails. Dad drove an old Ford pickup that all six of us children and our parents crowded into when we needed to travel. Rarely did we even want to go to town, for it was an exhausting, all-day trip.

> "Rarely did we even want to go to town, for it was an exhausting, all-day trip."

Dad raised corn and put up hay and moved haystacks with a team of big horses. He carried a lantern to the barn to milk cows by hand, then entertained us and the cats by squirting them a taste of fresh milk. Separated milk was kept on the porch to cool or drink fresh in warmer months. Cream was shipped on the mail route. We children would help churn cream into butter with an old crank on a jar or by shaking a quart jar. We made our own entertainment jumping, climbing or helping our parents. We read to the younger ones.

The local mail carrier, our link to the world, came by every Monday, Wednesday and Friday. Daddy later constructed a 32-volt battery charger to provide us with a luxury of radio which we used very sparingly for news.

The Rural Countryside

It was a rare treat to listen to the "Grand Ole Opry" on Saturday night; we gathered close to the radio. We were eating quietly around the breakfast table when news of the Pearl Harbor bombing came over the radio — and again when the first casualty, our neighbor's son, was announced on the radio.

The 32-volt system provided dim lights through a bare light bulb with a string attached, hanging suspended in the middle of each room. The washing machine was run by a gas motor. Many evenings were spent with the motor running to charge the batteries.

Our electricity and telephone came in the fall of 1953 while we youngsters were all in school. The older siblings were boarding in Thedford for high school. We came home on Friday night and looked around wide-eyed noticing the changes. The house was so bright.

A refrigerator was standing in the kitchen. We admired and tried every light switch to see how bright each room had become. Mom was looking forward to making ice cream without turning a crank on the freezer for hours like we had always done. We soon had running water and a real bathroom. It wasn't long before our mother had a toaster which she appreciated. Later came an electric washing machine, mixer and finally a dryer and dishwasher. Her treadle sewing machine was replaced with an electric model. She marveled in her last years when I bought a computer and learned to use it.

Our Grandma and Grandpa Paine had lived in a sod house with lamp lights where my older brother was born. We had witnessed progress from the age of pumping our own water or carrying it from the windmill in the yard. We learned to appreciate bright lights, running water, and an inside toilet and finally a real bathtub. The little old round galvanized bathtub became a fond memory, as did the outside privy, school kerosene lamps and so many old ways. I joined my family in being thankful to be "turned on."

Times dictated methods

by Millicent Anderson
Boscobel, Wisconsin

It's night. I am sitting here writing. I did not have to light a kerosene lamp to be able to see: I merely flipped a switch.

When I was born in 1925, very few people here in Grant County, Wisconsin, could merely flip a switch. A few who lived on major highways got power from Wisconsin Power and Light, and others powered lights, and maybe a radio, with their own light plants.

The industry in our area was farming with most farmers milking cows. A farmer had the number of cattle that he and his family could milk by hand.

Milk was cooled in springs or in tanks of water from a well. The water was usually pumped by a windmill. Most farmers hauled their milk to a cheese factory, but some used a separator and sold cream.

My great aunt sold cream. She stored her cream in a spring house. It seemed as though someone was stealing her cream. One night, upon hearing a noise, she lit her kerosene lantern and went to check. The door was slightly ajar, and in the spring house there was the thief — a raccoon lapping cream from a can.

Monday was wash day. One needed hot water. Most people had woodburning ranges, which they used for cooking and baking. My mother's range had a reservoir where water was heated. On washdays, she would also heat water on top of the range in a large metal boiler. Before my mother got her gas-powered washing machine with a wringer, she had to work very hard on Mondays. She set up tubs on a washbench. In one tub she would scrub the clothing on a washboard and then wring it by hand. Then she would rinse it twice and wring it twice. Then she hung it on the clothesline to dry. All that work really makes me appreciate my automatic washer and dryer today.

We had a woodburning furnace in our basement. Every fall my dad made sure that we had enough wood to last the winter. The heat came up through a large register in the floor and went through small registers into the upstairs bedrooms. When one stood on the big register, he would be very warm, but a short distance from the register he would get cool. (There was no circulating fan.)

However, with both the heat from the furnace and the heat from the cooking range, we were comfortable in winter, and my mother cooked and baked delightful treats.

In summer we had no choice but to use the range, and cooking and canning would make the house very hot. We kept our milk and butter in our basement, but one couldn't store raw meat there in the summer. That's why farmers butchered their cattle and pigs in the winter and most of the meat was canned.

And that's why my parents' first appliance purchase when they got REA in the late 1930s was a refrigerator.

Cold-weather farming rough

by William Stegner
Rhame, North Dakota

    I am a farmer-rancher and graduated from Rhame, North Dakota High School in 1941. I went to the agricultural college in Fargo for three years, served 18 months in the army during World War II and came home to take over my folks sheep ranch with my brother Harold as a partner.

    My folks started out living in an old sod house with a shallow well along side. It provided us with water using a hand pump - and had a large enough well-hole diameter where my mother would use a rope to put whole milk down in the water to keep it fresh.

    Then we skimmed the cream off the top. There was a hand crank separator to get the pure cream to sell and use on the table. We milked using a kerosene lantern which occasionally the wind would blow the flame out.

    Mother used a combination coal-wood stove with a reservoir that provided her with warm water.

    On the sunny side of the sod house, there were bumble bees which drilled their houses in the sod, and us kids would poke sharp sticks in the holes thinking we could kill the bees.

    Finally, dad built a new two-story house, the first in the area to have running water. A well was drilled along side of the house with a 55-gallon barrel installed on the second floor, and by pulling a lever on the hand pump the water would flow into the 55-gallon barrel. We knew when it was full when the water ran out the overflow pipe. This gave us one of the first water toilets.

    In order to start our loaded tractor without electricity we built a garage into the large dirt bank — covered it with planks and boards — then put about two feet of dirt on top, with a tight door. We parked the tractor inside, and were able to start it in all kinds of cold weather.

Our first electricity was a 32-volt wind charger, which allowed us to have our first refrigerator and water pump for water pressure inside the house. Prior to the 32-volt system, we used a gasoline lamp which had two mantels. The miller moths would fly into them and cause their light to go out.

Dad heated our first house with propane and because it was not insulated — frost would form on the outside walls.

We purchased a IHC Crawler® with a Bucyrus Erie® loader. It was terribly hard starting in the cold weather. I remember my mother became very sick and our neighbor walked through two miles of deep snow to help us get that diesel started. He put a bunch of gasoline soaked sacks under the engine and I thought the whole thing was going to burn up. Luckily, it didn't catch fire and we finally got it started, and plowed a road into our house where mother was put into a small hospital.

As the years went by, my banker helped us build a brand new house of 3,500 square feet. I hired a good carpenter who did an excellent job of building it. It's all electric, with forced air and ceiling heat. There is a beautiful waterfall with colored lights placed into the rock. The doors are recessed, so the wind will not blow them off. It has a large built-in two-car garage, well insulated, so it hardly ever freezes in it without any heat.

We also built a large cattle feed mill which is all electric, and the electric augers and hammermill all shut off automatically when the bins get full.

Our electricity heated home gives us warm, summer-weather living even when it's 40 below outside. And the electric air conditioning allows nice cool living during the hottest summer days.

Electricity brought evolution

by Gib Morfeld
Leigh, Nebraska

  I remember June 11, 1941 as the day in which my life as well as the lives of my family and neighbors were changed forever, a day in which we stepped into a new world when the REA energized the new line south of Stanton. I remember running a mile and a half to tell my father, who was cultivating corn. He worked for months to help get the 50 signatures required by the power district for them to build the line.

  We watched with anticipation as the lines were built and our home and barns wired. A new Philco® refrigerator stood in our kitchen waiting to save those many trips to the cellar. Light bulbs calmly hung waiting for the magic that would make them beacons in a dark countryside. When night fell that June day, the lamps stood as silent testimony to the dying days of an era of hard work, and the birth of a new life. The lanterns we carried around the farm gave way to a single large bulb in the yardlight and a bulb in each outbuilding that only required the flip of a switch. We used these switches often because our parents made sure we did not exceed the 50 kwh for $2.50 per month. This new-found luxury was not to be wasted.

  With ice in the refrigerator and ice cream without having to turn a crank or crack ice, trips to the cellar with milk, butter and leftovers ended. Food safety had a new meaning.

  We soon felt soft moving fan air in our bedrooms. Radio connected us to the outside world. We no longer had to fill the reservoir and boiler for wash day. The putt-putt of the Maytag® motor no longer greeted us on our way home from school.

  Eventually, the crank on the cream separator would stand idle in the corner, and the milk pail and stool would find a place with the lantern. We would no longer have to carry

The Rural Countryside

cobs or pick them up out of the hog yard or haul wood or ashes. Mom could cook, bake, can, iron and wash clothes without having to heat a room already 90 degrees. Often, the laundry chore was moved into the house from the wash house.

Although the immediate benefits of the lights came to the home, it was not long before production of the commodities that provided income for family living expenses also benefitted. With controlled lighting, the production of eggs vastly increased, as did poultry production, with the use of electric brooders to replace fire-prone oil ones.

Hand tools became electric driven, that forge was replaced by the welder, the scoop shovel by the auger, the hand pump by the air compressor. The romantic windmill stood still and it was not long before our rural schools and churches would be brightened by this new convenience.

Because electricity came to rural Nebraska, the use of forced air and grain handling equipment ended the days of cutting, shocking and threshing small grain in the summer heat. The sound of the ears of corn hitting the bang board on wet, frosty mornings, and the neighbors getting together to shell corn in zero-degree weather on a windy hill, would be replaced by air-conditioned and heated combines rolling through those same fields a couple of decades later. Each time the grain was handled it was done with a heavy iron scoop shovel in the hands of a farmer, his back furnishing the power.

I grew up in the days of Senator George Norris, the man who convinced Congress that without federal aid, no one would develop electric lines in rural areas because of the cost and meager investment returns. True, the initial returns were meager, but look at what rural Nebraska has become. Now, power companies would love to get their hands on the results.

Every single rural development in place today was influenced by those first copper wires that were strung to make life easier in rural homes.

Chapter 4

A Woman's Work

Wash day was an all-day job before electricity arrived.

Circle of Brightness

For the women of the farms

In his autobiography "*Fighting Liberal*" George W. Norris admitted that it was there on his Ohio farm where he "lost all fear of poverty," and that he learned to live "most simply" and "learned to get a great joy out of work."

There were few comforts in that home," George said about his childhood. "The loft, uncarpeted, unfinished, unheated where I slept, developed wide cracks between the shingles, and there were nights when I could see the stars twinkling in the Earth's celestial blanket. There were nights in the winter when the snow sifted in, too, laying a white blanket over my bed."

So George Norris was raised as most rural Americans — without electricity — and a nation of six million farmers continued their toilsome existence, much the same as their pioneer ancestors.

Farmers were slaves to daylight and horse power and only one out of ten farms in the United States had electrical service. While some farmers already enjoyed the benefits of gasoline-powered tractors and other machinery, their wives rarely had such luxuries.

The women often milked cows alongside their husbands by kerosene lanterns, then kept house with a monotonous routine involving cooking, cleaning and raising children. Much of their day involved fetching, hauling and heating water. One government agency estimated farm wives spent 20 more days per year washing clothes than a city wife using an electric washing machine.

This time-consuming task often discouraged cleanliness. If the wife wavered at any point in her daily routine, the lapse could be disastrous. With no running water, outhouses were often the source for health concerns such as hookworm. Food storage and malnutrition were also problems due to the unavailability of a variety of food groups and modern storage techniques.

A 1919 report by the United States Department of Agriculture, estimated rural families spent more than 10 hours per week pumping water and carrying it from the source to the kitchen.

By the 1920s, most politicians knew of the concerns of a rural countryside, but few dared to explore the thorny issue of how to remedy the problem. Senator Norris didn't have the answers, but he continued to look for a solution. He knew how important electricity could be to a nation's farmers and maybe more importantly — their wives.

What Others Said About Rural Living:
"The burden of the hardships falls most heavily on the farmer's wife than on the farmer himself. Her life is the more monotonous and more isolated, no matter what the wealth or poverty of the family may be . . . ."
— *Country Life Commission Report — 1909*

In George's Words: "I had seen first-hand the grim drudgery and grind which had been the common lot of eight generations of American farm women. I had seen the tallow candle in my own home, followed by the coal-oil lamp. I knew what it was to take care of the farm chores by the flickering, undependable light of the lantern in the mud and cold rains of the fall, and the snow and icy winds of winter. I had seen the cities gradually acquire a night as light as day.

Why shouldn't I have been interested in the emancipation of hundreds of thousands of farm women?"

In the words of the people, here are the stories of "A Woman's Work" . . . .

Circle of Brightness

Farm wife's work unending

by Eunice Kost
Washburn, North Dakota

No woman on this earth that lived without electricity on a farm and then got electricity could ever forget those remarkable days, when our lives changed from night to day with just the switch of a light bulb.

Monday was always wash day. I got up early and went outside to chop wood to start a fire in my kitchen coal range to heat water for washing clothes. I threw in a few dried cow chips to help it go faster. Then I burned coal on it. While it was starting to burn, I went to the basement to get my big copper boiler and filled it with water to heat to wash clothes. I carried water with two pails from the well until the boiler was full. Sometimes I had to pump the water out of the well if the wind wasn't blowing so the windmill could pump it. When the water boiled, I skimmed rust, hard water film, or scum off the top of the boiler before washing clothes by hand in the tub and scrub board.

Then I pumped more water and carried it to the kitchen — quite a ways from the water well. You had to pump and carry enough to rinse the clothing twice in water that had bluing paint to make the white clothing appear whiter.

I washed all my family clothing on the board with my homemade lye soap. No detergents in those days.

Freshly washed laundry was hung on the wash line outside in the bitter cold to dry. Of course it would freeze stiff as a board and sometimes tore when I took them from the wash line. Then I took them into the house to finish drying. The kitchen was about 12-by-16 feet with wash lines strung from end to end.

Men always wore long heavy underwear and those were

A Woman's Work

hard to wash and hard to dry.

Washing took all day Monday. Then came the next job of ironing — which took almost all day Tuesday. Clothing was stiff so everything that was washed was also ironed — towels, dish towels and all other clothing. I even ironed my baby diapers to make them feel softer.

It was tedious trying to keep the coal range red hot to keep the sad irons hot. No wonder they called them sad irons. Trying to keep the irons hot was a job in itself. I barely weighed 100-pounds at that time but the work never changed. I cried many times until I got done ironing.

There was hardly time to play and cuddle the little babies and children in the house.

I did my house work just like my mother and grandmother before me. No short cuts.

> **"There was hardly time to play and cuddle the little babies and children in the house."**

Those seven-pound wedge irons got heavy after hours of ironing. It was hard keeping them clean, while filling the range with coal. In those days every mother and child wore white shirts to church on Sundays or to funerals or other functions. Many times coal slack got on the iron and wasn't noticed until it hit the white shirt, which had to be washed again.

The same coal stove that heated the laundry water was used to cook meals during the harvest. The days were 18-20 hours long. Firing up the stove to cook for harvesters not only heated up a house that was already hot but made it miserable. In those days the harvesters were threshers — many of them. Food had to be baked. Bread and cookies were baked every day to serve to the hungry men. That was done by young brides. I don't know how I did it to this day, but thank God I came from a big family and I helped my mother cook for 30 people, five

Circle of Brightness

times a day at threshing time. When I look at young farm women today who don't even know how to bake bread and churn butter, I wonder how they would have survived in my day.

There was no refrigeration, so food was preserved as soon as it was harvested. I used to can more than 2,000 jars of food every year with the hot water bath, 3 1/2 hours a boiler full at a time — everything from meats to vegetables and sauces. There was hardly a jar that ever spoiled. I could go to my basement and find any kind of food I wanted.

It seemed like every day I wanted to can, the wind wouldn't blow so I had to pump the water by hand from the well and lug it to the kitchen for canning. My back nearly breaks just thinking of all that hard work. No wonder I had to have spine surgery in 1947.

Being pregnant in pre-electricity day was especially taxing. We had a little round galvanized tub that was used for washing clothing. It was also used on Saturday nights for the entire family to bathe in, in the middle of the kitchen floor. Every member in the same water. From youngest to oldest. When I was getting larger, I could hardly get out of the tub. I'd get stuck and no one was around to help me get out.

> **"Being pregnant in pre-electricity days was especially taxing."**

In those days, doctors delivered babies in the home. When the call went out for the doctor to come, he gave strict orders to heat up the fires and get boiling water. Doc said, "so he could sterilize his utensils." While Doc was doing his business, husband held a kerosene lamp over mom so Doc could see where to snip. We lived in a very cold house, using a coal range and a big coal heater - two stoves in three rooms

A Woman's Work

trying to keep warm. What a cold place to deliver a new baby.

On cold nights, we stayed up until 1 a.m. to keep the stoves fired up then we'd go to bed. By morning the water froze in the stove reservoir and the drinking water pail was frozen solid. It was so cold, the little baby's hand would be blue every morning. If we dropped a wet diaper on the floor at night it froze there, and would take nearly all day the next to get it thawed off the floor. I still wonder why my little boys didn't freeze to death in the old house.

I used to run to the outside toilet when it was 40 to 50-degrees below zero. That was regular temperatures — not wind-chill factors. When I sat on that cold seat, I nearly forgot what I wanted out there.

I helped milk the cows by hand and helped separate all the milk from the cream by hand. My arms felt like they would just drop off by the time I finished separating and feeding calves. I also had to clean all those kerosene lamp chimneys and globes from the lanterns. Those crusty things were always smoked up.

I cried when the electricity came on. It was just too good to be true. Just think! A warm home for my babies! We couldn't afford running water for two more years, but when we got it, that also was a miracle for a farm wife.

It was just wonderful to know we could live like our city sisters in a warm home. We could read at night.

Nothing on this earth — money spent to put man on the moon nor government aid programs — nothing has changed the standard of living in rural areas as much as the coming of electricity.

Mom didn't enjoy rustic

by Allen H. Williams
Cedarville, Michigan

With the Rural Electrification Act of 1936, electricity came to the remote area of Michigan's Upper Peninsula where I now live. Power on the mainland came first, but power out to the islands just offshore in Lake Huron took a little longer. My grandfather first set up a tent, and then a cabin on one of these islands as a summer retreat from the allergies that plagued him where we lived in the Lower Peninsula.

Mom and Dad took us five kids to the island each summer as well. We lived in a small frame cabin with a tent and outhouse out back. The REA had the cable out to the island, but not to our house yet.

Though we kids loved the rustic living, coping with the lack of power played an important part of each day for Dad and especially Mom.

Before we were out of bed, my father would have the little wood stove warming up the kitchen. My mother would cook breakfast on the kerosene stove and after the meal the rest of us turned to our chores. Kerosene lamps had to be filled, the stove tank topped off (woe to us if mom ran out of kerosene in the middle of cooking a meal), wood boxes filled, water buckets filled from the clear blue waters of Lake Huron, and the outhouse swept free of spiders.

> **"Woe to us if mom ran out of kerosene in the middle of cooking a meal."**

About every third day, we went into town to buy groceries, kerosene and ice. In the early days my grandfather rowed the three miles and back in one of our long white rowboats with a wineglass shaped stem. We had outboard motors to make the trip.

The same wash tub we took our weekly baths in was placed in the boat for the trip. In town, we would place two large cakes of ice in the round tub and then break up a third to fit around the sides. With an icebox, there was no such thing as getting a cold soda out of the refrigerator. Only essentials could be kept in the precious little space of the semi-cold box. There were also strict rules about who could open the icebox and when, so as to stretch the ice as long as possible.

Nature's refrigerator was the lake at our doorstep.

We pumped cold lake water into the bilge of the sailboat to cool Dad's beer cans.

Fresh fish did not need refrigerating and we went out and caught dinner most afternoons. After dinner, we continued to enjoy the long summer daylight but we kids went to bed shortly after sunset.

The adults and older kids stayed up playing cards or board games by the light of kerosene lamps. Trimming of wicks was an ongoing process.

In the mid 1950s, we moved down from the island into a bigger log cabin — that the previous owner had electrified.

This move was in no small part to appease my mother whom I know did not enjoy rustic living as much as her five kids did.

Her next goal was indoor plumbing.

Circle of Brightness

Electricity lights new world

by Lois Moore Duvall
Westminster, South Carolina

In the 1940s, it was like moving into a new world when the lights came in our country home seven miles from town.

Before we got the power, we had a wash place down at the branch. There we heated water in a big black wash pot, dipped it out into a tub to wash the clothes, then filled it up again from the spout in the branch to boil them. We battled them on the battling bench with a battling stick before boiling them.

We used kerosene lamps at night and sometimes we would run out of oil before market day, and my mother would make a torch with pine splinters laid on an old bread pan for a light.

> **"We battled them on the battling bench with a battling stick before boiling them."**

After the lights came the ringer-type washing machine, refrigerator, electric stove and too many other things to mention. Not all at one time, but as we could afford them.

Blue Ridge Electric Co-op has brought us and many others a long, long way since those days.

The joy of electric ironing

by Mrs. James Gaulden
Rock Hill, South Carolina

It was indeed a great day when we got electricity. Life changed a great deal for the better.

I don't remember exactly the year, but sometime after World War II.

We did get an iron fairly soon after we got electricity. What a joy to be able to plug it in instead of using a flat that was heated on a wood stove. You could never get the iron exactly the right temperature on the stove and you were forever coming up with scorched white clothes. I guess other things got scorched but it did not show as much.

> "...you were forever coming up with scorched white clothes."

I remember the bright overhead lights that you pulled a chain and everything was brighter than it had ever been before. The neighborhood seemed to take on a different look when you could look out the windows and see the electric lights brightly burning. There was no more washing smutty lampshades on Saturday, and trying to do homework by lamplight.

We had very few appliances for a good many years, but we were happy to just have bright lights — and the electric iron.

A girl raised on lantern light

by Elizabeth Mahocker
Cable, Wisconsin

When I was a girl we had lanterns, lamps, outhouses, wood-burning stoves, a hand pump in the yard for our water supply, and a tin tub for bathing. In the winter, bathing was done next to the wood stove for warmth.

The lanterns were our light source for doing the barn chores and other jobs before light and after dark. Milking was done by hand. Windmills pumped water for the animals. Spells of no wind meant water had to be pumped. Kerosene lamps lit the house. Each day the lantern and lamp chimneys needed washing. Especially if the wicks weren't trimmed properly, then the chimneys would be blackened. They also had to be filled with kerosene.

Outhouses were our toilet facilities with a chamber pot for night use, especially in the winter. The fire in the kitchen stove was always kept filled with wood, summer and winter for cooking and baking.

Most cookstoves had a reservoir that was kept filled with water. The water was always warm and served many purposes. There was always a teakettle on the stove for instant hot water for cooking and doing dishes.

The night before the laundry was going to be done the boiler was put on the kitchen stove and filled with buckets of water. The hot water was transferred to the wash tubs setting on homemade wooden benches. The clothes were scrubbed on a wash board with bars of P and G® or Fels-Naptha® soap and a stiff brush. The laundry was wrung by hand. After rinsing in tubs of fresh water, clothes were wrung by hand again and hung outside on clothes lines to dry.

Many things had to be ironed. The irons were heated on

> "The first laundry improvement I remember was a wringer that clamped on the side of the tub."

the kitchen stove. The first laundry improvement I remember was a wringer that clamped on the side of the tub. You turned a handle and guided the clothes into rubber rollers to eject the water. My three brothers and I turned that handle a lot. Life got a lot easier when we got a Delco® engine which operated a few light bulbs and a washing machine.

Perishable food was kept cool in an ice box. Large blocks of ice were put in separate compartments and as it melted the water was caught in a receptacle underneath. Basements were cool and fresh vegetables could be stored there, some buried in the sand. Lots of home canning was done. I didn't have much experience with indoor plumbing until I went to school. Life back then was hard but good.

Now, I wouldn't want to be without my lights by touching a switch, my refrigerator, washer and dryer, all of my electric appliances, bathroom with flush toilet, all of that hot and cold running water, and instant on and off stove and heat by turning a thermostat.

I could do without my TV, but the microwave sure is handy!

Circle of Brightness

Ode to Michigan farm wife

by Ethel Bernhardt
Wayne, Michigan

Do I remember what it was like before electricity came to my home? Did electricity change things in rural America? It sure did for my mother!

What a flood of memories come with these questions. It was 1937, I was 17 and just finished high school. My two brothers were older and a sister was already married. My home was south of Onaway, Michigan, about 3/4 of a mile. Of course the cities or towns had electricity, but not until the R.E.A. came, did country people have that power.

Oh, those kerosene lamps were something! In the living room was an Aladdin® lamp, which had to be filled with kerosene almost every day. In the lamp was a mantle, white, very thin. Mom would remove the chimney, light the wick with a match then we had to wait about a minute for it to flare up and light the room. It did make a terrific light. The folks did a lot of reading and I had homework every night.

Kerosene lamps were used in the kitchen and other rooms. I remember one with a reflection that hung on the wall. Chimneys had to be washed. Wicks had to be trimmed. They had to be filled often with kerosene.

One evening after dark, dad came running through the house. He grabbed the Aladdin® lamp and threw it out into the snow. He had accidently filled it with gasoline. The lamp did not explode, but we were fortunate not to be hurt or burned..

Now let's do the laundry. My mother — bless her heart — kept her family clean but how in the world she did it, I'll never understand. On wash day she unfolded a wringer, a wooden thing with space on each side for a tub and a wringer in the middle to turn by hand.

A copper boiler was filled with water and heated on the kitchen range. One tub was filled for rinsing, and one for scrubbing. My dear mother scrubbed all those clothes on a washboard and boiled some too.

Where did the water come from? We had a well out in the yard, and a cistern full of rainwater. It was pumped by hand and carried in, but dad and the boys often helped. Clothes were pinned on a clothesline in warm weather.

In bad weather, I hated wash day. I came home from school to see clothes on lines throughout the house. Anyway, the house was always warm. It was heated with the old coal and wood heater and the kitchen range which burned wood.

Oh, what about ironing? On that range, mama set three sad irons. With a wood handle she used one iron until it became cool, then trade it for a hot one, back and forth.

My mother was a fabulous seamstress making all our dresses on a Minnesota treadle machine. She also make a quilt or apron — whatever was needed.

Did we have any fun? Sister and I studied piano, we always had a good one and always had lots of good books. Brother had a guitar and we had great times with music.

I have been thinking about the ladies work. What about the men and boys? If there were cattle, light was by lantern. Cows were milked by hand. Milk was carried into the kitchen and strained into the big bowl of the cream separator. Then someone had to turn the handle. I did that many times.

Refrigeration? We carried the cream down into the cellar to cool. Milk too. We sold some cream, Mama made cottage cheese. Was it good! She also made all the family's butter. Dad wouldn't eat "store" butter.

All of the jobs were made easier when electricity became available. But most important — light. God said, "Let there be Light." It truly was an act of God. Anyway it seemed so to us and I'm sure it was for my mother.

'Sawdust Girl's' job changed

by Winifred White Blodgett
Hamilton, Montana

I was just out of high school when we got electricity. I lived in the back of a country store and post office with my mother, who was the postmistress and owner of the store.

We had an ice house where blocks of ice, taken from the river, were stored. They were kept frozen by being covered with lots of sawdust.

It had been one of my jobs to get into this sawdust, help get the ice out and clean it off for use in the ice boxes in the store. What a relief to have electric refrigerators and places to keep and display meat, pop and even ice cream!

On January 14, 1938 the first electric lines of the Ravalli County Electric Cooperative were energized, serving 118 miles and 271 members in the upper Bitterroot Valley of southwestern Montana, possibly the first in the state.

> "It had been one of my jobs to get into this sawdust, help get the ice out and clean it off for use in the ice boxes in the store."

We also sold gas and kerosene pumped by hand from underground tanks. I can remember that big change when we were able to get electric pumps for those. Pumping gas and kerosene had also been one of my jobs.

I remember well how excited and thrilled we were when we finally received rural electricity and could put away all the kerosene lamps and gas lights.

It was hard to believe that there was so much good light

to read and study by with just a flick of a switch or pull of a cord. No more glass lamp shades to wash to get rid of the soot when the lamps smoked.

We were also able to do away with the big black folding bathtub — I think it must have been rubberized material of some kind — and the boiler on the wood stove to heat water for baths. At last we could have an electric water heater, electric water pump and a real bathtub! What luxury!

Before we had that, I can remember my mother using the bath water after baths to mop the floors. We didn't waste warm water, believe me! We didn't waste any water because it had to be pumped out of a well by hand. Often times the pump had to be primed, and if that didn't work, water had to be carried from the river.

Everything was different and although we didn't get to change things all at once because of finances, what a great day it was when we could get rid of the washing machine that was turned by hand. We got an electric one with a wringer that even worked by merely turning a handle, and what a difference to receive hot water from a faucet instead of the boiler on the wood stove.

Hooray!

Yes, the coming of power in 1938 to our area surely changed the way of life for many.

Circle of Brightness

Mother sacrificed career

by Luke B. Hart
Whitmire, South Carolina

For a farm family of eight living on Pea Ridge in Union County, 1941 was the year the lights came on. This was a happy year for the mother of six boys reared on this 150-acre farm.

Cooking on that old wood stove for six hungry boys was an enormous feat for a mother who had given up a teaching career for family life at the age of 31. Before electricity, the family had always burned wood for heat, cooked on a wood stove, and worked by candlelight or kerosene lanterns.

Education was important to our parents and all studying was done by kerosene lamps. Although it was hard to study by this light, four of the six boys received college degrees.

Of course, chores always played a big part in keeping a farm going. Cutting wood for heat and cooking had to be completed in the winter so fields could be plowed during the summer.

When rural electricity came to Pea Ridge, farm life became easier — especially for mother. Modernizations of kitchens, bathrooms, and the warm glow of lights made our house a better place to live.

REA: Change for women

by Doris Hulse
Free Soil, Michigan

When I think of the years my family spent without electricity, it stirs many memories of an entirely different lifestyle: clothes washings done with tubs and wash boards, ironings done with sad irons heated on a coal and wood range. Sad irons were aptly named, for if a piece of soot from the top of the range adhered to the iron, a long, black streak adorned your blouse or husband's white shirt.

REA — and now Western Michigan Electric — brought electricity to our area in February of 1949 and ended the years of reading by lamp light or listening sparingly to a battery-powered radio.

It also ended a great amount of drudgery for rural women. I can only imagine what a boon it was to the farmers.

My memories mainly reflect what a remarkable and memorable change it brought to the women of our area. Just the fact that we were able to have running water and indoor plumbing was a joy.

Those days of hardship are gone forever and I, for one, am grateful to our rural electric providers for ending them.

Circle of Brightness

Letter recalls flatiron days

by Norma Te Selle Prophet
Firth, Nebraska

I have been on this planet earth for three-fourths of a century, and have seen many changes. One event, which stands out in my memory as one of the most important and one that benefitted so many people, was the implementation of rural electrification.

"Mom said they have lights now and I know that she is really happy about it, there isn't anything like electricity."

To an outsider, this excerpt from a letter I had written to an aunt on Oct. 2, 1948, could have been interpreted as a mere statement, but to my parents, who were living in rural Firth, having electricity on the farm was a long awaited event!

I was a privileged farm child, I had the privilege of helping my mother with the weekly ironing. This task was accomplished by making frequent trips between the kitchen wood burning cook stove and the ironing board. While using one of the flat irons, two more were heating on the stove. And all the while, wood or cobs were stuffed into the jaws of the cook stove. The stove heated the irons and the house.

In those days, we ironed everything, I think, and the wearing apparel was starched and dampened. Wash-and-wear material was not invented. Flatirons were in vogue.

When I was in high school, one of my Saturday chores was to clean and trim the wick on the kerosene lamp and fill it. When centered on the kitchen table, it provided sufficient light during our evening meal. I also sat at the same table to do my school homework.

My mother always had a big garden and did a lot of canning. Canning was an all-day undertaking; a steady heat had to be maintained in the cook stove so that the canner would heat properly. She also canned beef and chickens. Canning was the only way to preserve meat for future use.

When my folks' farm was wired for electricity in 1948, my husband was a college student at Stillwater, Oklahoma. When we visited them, it was great to see the farmstead lit up with a yard light at night.

One of the first appliances my folks purchased was a small chest-type freezer. My mother also acquired an electric cooking pot, a portable electric roaster/oven and no doubt, an electric iron! My folks were definitely uptown.

Although there are now wash-and-wear clothes on the market, I still iron the shirts, the trousers and the dresses, and it is easier with an electric steam iron in an air-conditioned home.

Among my keepsakes from the farm of the olden days, I have the three old flatirons, the old kerosene lamp, and even my mother's electric cooking pot.

Having electricity for the rural farm families, in my opinion, was one of the greatest accomplishments during the 20th century.

I never want to return to those flatiron days.

I'll take 'now' over 'then'

by Doris Kuffer
Shullsburg, Wisconsin

To light the house, we used candles, kerosene lanterns and lamps, mantel lights, and then electricity.

It was early to bed, early to rise.

There wasn't running water, you pumped it by hand and filled boilers, then heated the water on the wood stove, dipped it out into two galvanized tubs to wash all the clothes. Then you had to wring clothes by hand and hang them outside on a clothes line with clothes pins.

Now we have several cycles on an electric washer and temperature controls for the water. Also electric dryers with a variety of settings.

To iron, we put the base on a cook stove to heat, and set them in different areas of heat, according to what you were ironing.

Now we have an iron with several temperatures plus steam. Where before you would sprinkle the clothes before ironing.

We depended on a windmill to supply water.

Now electric pumps pump water and electric water heaters heat it.

People bathed once a week, in those galvanized tubs in water heated on the boiler on wood cook stove.

We milked cows with a three legged stool and bucket between their legs. We sometimes milked by kerosene light and had to shovel the barn clean of manure.

Now with electricity, the milk is all pumped directly to a tank. Now we just push a button on the electric switch to "shovel" the barn.

We hung our area rugs on a fence or clothes line and

beat them.

Now we have electric vacuum cleaners, brooms, scrubbers, shampooers, to take care of floors.

Baking was mixing up everything by hand in a big bowl with a wooden spoon.

Now we have electric mixers, blenders, choppers, knives, and an electric stove.

Many families played piano or an instrument of some kind to entertain. To play records, 78s, you wound up the Victrola. Now we have VCR's, TV's, cassette players, CDs — all with a remote control.

People played a lot of checkers and cards, did a lot of hand-sewing, also made clothes and other things for the house on a treadle sewing machine. Now we have electric sewing machines that can do most everything.

Washing dishes was done in two pans of water. One to wash and one to rinse.

Now you put them in an electric dishwasher.

Saws, drills, sanders, screwdrivers, planers, and hammers — everything used to be by hand.

Now we have electric screwdrivers, drills, staples, saws, sanders, drills and saw mills.

We butchered quite often as they didn't have freezers and we canned a lot of meat.

Now we have freezers.

We went from iceboxes to refrigerators, with running water and automatic ice makers.

We went from rags to curling irons in an oil lamp to electric curling irons of all sizes and electric hair dryers.

I'll take "now" over "back then."

Circle of Brightness

Mama, REA made difference

by G. L. Williams
Aiken, South Carolina

In the country, outside Graniteville, South Carolina, Mama, Daddy, and seven children lived in a three-room farmhouse, complete with wood shutters for windows, a fireplace for heat, oil lamps to do our lessons by and a wood stove for mama to cook on. The house was completely dark, except for the flickers from the fire and an oil lamp, that mama kept on the mantel for extra light in the room.

Sunday morning. December 7, 1941. All of us were sitting around the fire, talking when we heard a horn blow out front. Excited, we all ran to see who had come by. It was a neighbor who came to tell us that the Japanese had bombed Pearl Harbor. Not having electricity, we were not able to have a radio, so we had to depend on someone stopping by to tell the news. The news was exciting and scary to me, I was so young and could visualize the Japanese soldiers coming to shoot us.

February of the next year, we moved in to a new four-room house. Mama was so excited, but she still wasn't satisfied. We still had no electricity.

Somehow Mama heard that we might be able to get REA to run power to the house if enough people signed up in our area.

Being a fighter, Mama quickly threw all of us kids in the old Plymouth, and away we went down the dirt road to get our petitions signed. I'll

> **"Mama quickly threw all of us kids in the old Plymouth, and away we went down the dirt road to get our petitions signed."**

never forget all us kids, heads hanging out the car windows, excited that we were going to get electric lights. Mama pleaded with all the neighbors to sign the petition, but some were reluctant, because they were afraid that they would not be able to afford the light bill every month, but she didn't give up.

We kept riding and me, being the oldest of the seven children, sat in the front seat. One day, my sister started screaming that my brothers Nelson and Milton were gone from the back seat. Looking back, we saw the door open and found we had lost the two boys.

Mama turned the old car around, headed back down the road and saw the boys lying in the road crying, scratched but otherwise unhurt.

Finally, we got enough people to sign and mama got a man from Graniteville to wire the house. I remember all of us huddled around the REA man and watched him, praying, that he would pass the inspection.

Thank God for mama and the REA. That was a happy day for us. He passed the inspection and now we had lights. We could do our lessons and hopefully we could get a new stove for mama to cook on, and a refrigerator so we could have ice for tea, and a radio to hear all the news.

Yes, Mama and the REA made a big difference in our lives and I appreciate both of them very much.

Chapter 5

Political Shadows

Everyone was interested in the electrification plans.

Circle of Brightness

The politics of electricity

When a three-year old George Norris found out his brother John was killed during the Civil War, he sensed his mother's devastation. He rummaged through a closet, came out with a buggy whip and reportedly yelled, "I'll whip that rebel that shot John."

Aware of worldly injustices at a young age, George became a motivated school student training for a career of correcting injustice. He attended Mount Carmel district school not far from his Ohio home, studied almost every night by the light of a candle. In 1877, at age 16, he enrolled in Baldwin University and earned extra money teaching. In 1879, he enrolled at the Northern Indiana Normal School and Business Institute — later known as Valparaiso University — and returned home four years later with a law degree.

After a brief attempt to set up a law practice in Washington, Norris settled in eastern Nebraska where his mother owned 80 acres of land. In 1885, Norris set up a law office with a former classmate but that partnership dissolved after a couple of months. Norris sold the 80 acres and moved west to Beaver City, Nebraska, where he established a thriving practice and met his wife Pluma Lashley in 1890, but she died in childbirth, delivering the last of their three daughters.

Forever aware of social inequity, Norris ran unsuccessfully for Furnas County prosecuting attorney. He won election to the same post in 1892 and was elected judge of the 14th Nebraska District Court in 1895.

In 1902 he was elected to his first of five terms to the House of Representatives and elected to the Senate in 1912.

In the Senate, Norris developed a reputation as a protector of farm rights and natural resources. He also kept a watchful eye on big business and power trusts.

State, regional and nationwide committees were formed

to study the possibilities of rural electrification and its potential impact on agriculture

Throughout the 1920s, Norris battled the biggest bully on the block when he took on the omnipotent "Power Trust." He constantly emphasized the need for lower electric rates and more extensive service.

Many reports showed that rural electrification was practical but the nation's power trust failed to act — and why would it? Under the system in place, it made no economic sense for neither the power companies nor for the rural consumer. The farm market was too small to warrant additional expenses of building costly rural power lines. Likewise, it made no sense for the farmer to pay the extra money (sometimes as much as $6,000) to foot the entire cost of getting hooked up. Even if he could afford the hookup charge, he wasn't going to be able to pay for the high cost charged to him for electric power.

It became a matter then of social responsibility and that was the opening for the "Gentle Knight of American Democracy" to wedge his political sword.

What Theodore Roosevelt said about power trusts: "It is the obvious duty of the Government to call the attention of farmers to the growing monopolization of water power. The farmers above all should have that power, on reasonable terms, for cheap transportation, for lighting their homes, and for innumerable uses in the daily tasks on the farm."

In George's Words: "The power trust is the greatest monopolistic corporation that has been organized for private greed."

In the words of the people, here are their stories about the "Political Shadows"....

Circle of Brightness

Strong foundation required

by Lelah M. Gilbert
Angola, Indiana

Yes, I remember exactly what I said when the lights came on.

Much excitement had been generated in the preparation of this great moment.

Everything of great value requires the laying of a good, sound foundation. In this foundation, there were many days spent securing easements for erecting high lines for transmission of electric power. Skilled labor for the building of high lines was an absolute necessity. Equipment needed by those who would do the actual building of lines was costly, so financial supply was another must.

The state of Indiana was among the first statewide organizations to incorporate within their statues the element of "Necessity of Convenience."

> "...there were many days spent securing easements for erecting highlines for transmission of electric power."

This foresight made possible the privilege to apply for financial aid from the National Rural Electric Corporation Association.

As these qualifications were met, Indiana was ready to go into action. Time moved along and our rural electric lines in Steuben County, Indiana were erected and ready to be energized.

We had a Delco® system powered by 16 batteries with a gasoline motor to keep the batteries charged. These batteries were located in the basement.

We had purchased an electric radio which had an adaptor and our radio was ready to operate from the electric highline. What an exciting time it was.

I cooled six bottles of milk each morning and set them in a pan of cold water on the basement floor. That way I had milk for our 18-month old baby Stanley (who grew up to become a member of the Board of Directors of our local Rural Electric Membership Corporation in Angola, Indiana).

Yes I remember very well where I was and what I said on The Day The Lights Came On.

I said "Praise the Lord, now we can buy a refrigerator."

I was in the basement getting a bottle of milk for our little Stanley.

FACT NOTE: Our first refrigerator was bought and delivered four days after the day the lights came on.

All the rural electric cooperatives all over the United States owe a vote of gratitude to a great pioneer, George W. Norris.

Circle of Brightness

A 'hog trough' explanation

by Dic I. Craven
Hastings, Michigan

I lived with my grandparents in 1936. We hand-milked a herd of cattle and goats, pumped water and carried it to them.

I came home from school one night and saw men attaching brace wires to the poles they set by hand that day. There were no cross arms on the poles, just single poles. Gramp convinced neighbors to sign up, since the REA would come out if a majority of them signed up. Gramp was given a brass-based, table lamp for his work. I still have the lamp.

A friend wired our house and barn for us. He used an old car with the rear seats taken out to carry his tools and equipment. He used No. 14 Romex, rubber covered wires with paper insulation, then covered with a black, coarse mesh casing. We used a 25-watt bulb in each room and two in the barn. We also had one outlet in each room. Each evening when we did chores, the lights would blink and go out for a few hours. Gramp asked an electrician friend why this happened.

He said, "If you had 50 or more hogs and a feed trough long enough for them, if you poured a five gallon bucket of slop in at the top end, the ones at the end would wonder why you called them." Our power was coming all the way from Burnips Corners, and it was generated by a diesel engine.

We noticed a neighbor had his lights on all day. He was in a small home with single bulb. We asked why he left his lights on all day. "They told me I wasn't using enough power for them to make out a bill for the minimum amount."

Today, we live on the same land, but are blessed with a large refrigerator, two freezers, a computer, washer, dryer, and many other conveniences. But we remember the days before there was any electricity.

Second 'Battle of Gettysburg'

by Wayne S. Cohick
Gettysburg, Pennsylvania

In the summer of 1941 as a 14-year-old farm boy, I watched the Adams Electric Coop lineman put his hot stick to the transformer to energize our farm buildings. I ran to the barn and made sure all the lights worked before he left. After milking the cows and doing my chores that evening, my shoes were clean because I had plenty of light to see where I was stepping.

After my parents bought a refrigerator, some of the highlights that I enjoyed most were having ice cubes in my glass of water and an electric motor to pump water for 25 cows and four horses.

In January that year, six miles away, my future father-in-law, R. Boyd McCullough — along with 50 or more neighbors — participated in the "battle of the poles" against the investor-owned utility (I.O.U.), which built lines one mile on each side of his farm, but would not build to his farm.

When the co-op started staking out a right-of-way, the investor-owned utility started to connect the two-mile section that they refused earlier. This group of farmers decided that the IOU would do it now. After the IOU dug holes and put in poles, the farmers back-filled holes and sawed off poles.

Their action ended up in the Cumberland County courts by March 19, 1941 resulting in eight days of hearings, and more than 1,000 pages of testimony.

The county judge ruled that it was in the hands of the Public Utilities Commission. Commission Chairman John Siggins ruled in favor of Adams Electric Co-op, also stating that the real point of the controversy was that electricity should be made available to the rural people.

Circle of Brightness

A visionary emerges

by Ella Michalek
Camp Douglas, Wisconsin

The time was the 1930s. The stock market had crashed a short time previously, and things were bleak. Unemployment was fifty percent and despair was everywhere. America was in a Dark Age — especially rural America. Without electricity the kerosene lamp cast its yellow glow. The wash-boiler on the wood-burning cook stove heated water for bath and laundry. The scrub board and hand wringer completed the laundry job. With no refrigerator or freezer, no radio and no indoor plumbing, it was sparse living in the home. Likewise in the barn. The kerosene lantern furnished light while the farmer hand-milked the cows. Water for cattle was hand pumped. This truly was the "Dark Ages" in the country.

The whole country was in desperate straits and something had to be done.

In order to get the economy to move forward, a dramatic change had to be made. As a start, power had to be expanded to the rural areas. This would allow farmers to produce more and better food for a higher standard of living for everyone. It would also permit businesses and plants to move into spacious quarters. This is what the country needed.

> **"The whole country was in desperate straits and something had to be done."**

And as often happens, when things are at their bleakest, a visionary emerges. It would be a person who was aware of the despair, but was in a position to do something about it. This man was George W. Norris, a United States Senator representing Nebraska. He had the foresight and the fortitude to rise above the gloom and propose legislation to make the necessary change.

Since the whole country had to be involved it needed to be a federal program. It was an awesome task he had before him. The vast stretches of America had to be wired. This meant a great deal of political maneuvering to pass the legislation, and also to appease the private power companies, who felt this would infringe upon their rights. Eventually the legislation passed with a great deal of jubilation from rural America.

There was elation everywhere. The Dark Age of America was over. Gone were the cook stoves, wash-boilers and kerosene lamps. They were relegated to attics and eventually antique stores. Farmers installed milking machines and milk coolers to increase their income. For potato and vegetable growers it meant irrigation for productive yields. It expanded agriculture to America's highest standard of living, not only to feed us, but to feed starving people all over the world.

Rural Electrification illuminated not only the countryside, but gave us a whole new lifestyle — a lifestyle that transcended from rural America to impact the whole world. Its reflection was apparent everywhere. People had freedom from drudgery to enjoy life.

Again, thanks to Senator George W. Norris, who had the vision of a Greater America, and pursued it to victory.

Circle of Brightness

Cooperative anxiousness

by Mrs. Joe Russell
Hickory Grove, South Carolina

Electricity was first supplied to areas around York County, South Carolina in the early 1900s with the building of the Catawaba River Dam in 1904. Private power companies supplied power to the surrounding towns, but it was years before rural areas would benefit from this luxury.

Our neighbors a couple miles down the road had electricity supplied to their houses by the power company. Seeing the things that electricity provided made us want it so much, but the power company did not supply power to rural areas. It was not cost-effective to run the poles and line required to supply power to sparsely populated areas.

York Electric Co-op (REA) provided electricity to rural areas where we lived in 1941. The farmers were so excited to get power that in many cases, they helped the REA cut the right of ways.

Once they built electric lines to our house, the first priority was to get lights. The lines were not hidden in the walls as they are now, but stapled on walls. One of the first electrical appliances that our family owned was a refrigerator. We were amazed that the Kelvinator® would not only keep our food cool, but a light would come on when the door opened!

Our lives changed slowly as we added more electrical appliances.

Growing up in rural York County in the 1930s was

> "The farmers were so excited to get power that in many cases, they helped the REA cut the right of ways."

extremely difficult, compared to today's lifestyle. Our homes were lighted by kerosene lamps, so bedtime came very early. We cooked and heated with a wood stove. It was a constant job to keep a supply of wood that had to be cut with axes and hand saws.

Food preservation was not easy. We kept milk cool by storing it in a spring. We purchased large blocks of ice from a store nearby. We stored the ice in a pit filled with saw dust. It would last only a few days.

A fire usually burned in our yard. This heated a large black wash pot filled with water. It supplied us with hot water for bathing and washing clothes. We pulled water from a hand-dug well in our back yard. Much of the time, we supplied water for these tasks by catching runoff from our roof when it rained. We bathed and washed clothes using large tin tubs. When we washed clothes on a scrub board, we hung them on a fence to dry.

Had the REA not been there to bring power to our areas, it likely would have been many more years before we benefitted from this luxury.

A few years ago, hurricane Hugo destroyed many power lines in our area. Many people were without power for days and even weeks. They realized that electricity is not a luxury, but a necessity. Those of us who can remember when it did not exist, it is still a luxury.

Circle of Brightness

Joined REA for $5

by John Zurian
Moquah, Wisconsin

In 1939 local merchants talked to a representative of the Ashland Power and Light Company. The Ashland Power survey found a six-mile stretch would cost $700 a mile with only two farm outlets per mile. That was too expensive for our rural community. Later that year members of the Bayfield County Fair Association met with a representative of the Douglas County REA, who agreed to sponsor the REA in Bayfield County.

Each community in Bayfield County held meetings and voted on this project. About 65 percent of the voters in the county sided with the REA. It cost each settler a $5 entrance fee and that is how the Bayfield County REA was born. Soon money from REA of Washington, D.C. was loaned and work began.

In the fall of 1941, holes for posts were being dug by hand, with poles and wires going up. Twelve men from our community took part in this project, including myself, were supervised by the Jump River REA Coop from Ladysmith, Wisconsin.

The project was cancelled with the bombing of Pearl Harbor, but started up again once the war was over.

Contractors from Phillips, Wisconsin did most of the wiring in our area. They were limited to only 75 pounds of copper wire per farm. They charged us $128 for the wiring of our farm. The only outlets we had were with a 1.5 kva transformer and a 60-amp entrance box in the kitchen.

On January 1945, electricity came to my dad's farm in Moquah, Wisconsin.

Officers got electricity first

by Richard Woody
Crawfordsville, Indiana

My father was secretary/treasurer of the Boone County REMC and became the first person in the nation to get electricity from the REA. The board president lived about a mile away and was supposed to have the first meter, but he wasn't home. Our next-door neighbors received his meter, then the president returned home and received the third and last meter set that day. That night the neighbors were invited to a party, and all enjoyed the lights. They were soon to get their own electricity.

When I started to school the country school had electricity, but the school and the Quaker Church across the road and the next farm house where the farmer had a dairy farm, was the end of the line. At home, we used kerosene lights and acetylene light system that we recharged every summer.

The lights came on May 22, 1936. I was driving a team of mules hoeing a field of corn. When I came to the house we had electric lights! I went from room to room and flipped the switches to see the lights come on.

Today in the house I live in, we have electric heating coils in the ceiling and a thermostat in each room. When a room is not in use we do not heat it. With electric washers, dryers, water heaters, dish washers and other electric appliances, times have changed immeasurably. People never really appreciate the electricity until the lights go out — especially in cold weather.

I have been a trouble caller and recently during a snow storm, a neighbor called to see if I could plow out the road as they had two small children. I reported the trouble and 90 minutes later, she had electricity.

Circle of Brightness

Political battle within family

by Jean (Wilson) Stickney
Custer, Michigan

Three days after my sixth birthday, the power came on at the home of my parents on June 8, 1938. To be hooked onto the high line marching down Miller's field across the road meant more than flipping a switch for a light. It was a victorious culmination of struggles and dreams for my father. Though his dreams are vivid in my memory, I was protected from or didn't understand his frustrating struggles.

Two names I heard often from the time I can remember anything were those of Sam Rayburn of Texas and George Norris of Nebraska, senators far away in Washington D.C. I only knew my father considered them the good guys, part of his hope for an electric company to bring power to our farm.

I would have been shocked to know my father's closest antagonist was his own father — my beloved grandfather — with whom my father farmed the family acreage. I am so thankful I was an adult with reasoning to understand how each man felt before I learned of their argument.

I refer to my father's memoirs in his own handwriting that tell of the two opposing views he faced while he worked on a feasibility study for a local cooperative. Referring to the Senators Rayburn and Norris he wrote:

"Power companies were saying farmers were too poor to pay a light bill (and so the venture) would not be profitable. But the senators knew that farm income from the use of

> "I would have been shocked to know my father's closest antagonist was his own father."

electric power would make farmers capable to pay a light bill."

In spite of his father's opinion, he wrote:

"My dad was sure that a government loan to a bunch of dumb farmers was only a way to let the government come into our lives and tell us what to do. He often urged me to stay at home, cultivate the corn, keep the old Delco® running for lights and leave the government out of our business."

The first appliance purchased for the farm, even before the milking machine, was an electric cooking range for my mother. I have heard her tell in reflection of waking in the night to go to the kitchen and rub her hands over her new stove to be sure she had not been dreaming. The stove was not altogether unselfish of my father for this meant one less stove for which he would have to cut wood.

My father would remain an REA board member until 1955 when he became a full-time employee of the coop in charge of field services until his retirement in 1973.

Herein is concrete evidence of the varied sights I saw June 8, 1938, the day the lights came on.

Sharecroppers turn to Co-op

by Katharine Rampey
Central, South Carolina

    The day the lights came on in 1952 was the day Blue Ridge Electric Co-op cast the deciding ballot in where my husband and I — along with his mother and our four children — planted our next crop in Pickens County, South Carolina. It is where we remain today.

    After moving from sharecrop farm to sharecrop farm all of our lives we wanted to purchase a farm of our own. With no collateral with which to apply for a loan, we decided to turn to the Farmers Home Administration. A home purchased through the FHA had to meet certain building qualifications. Finally we found one for sale that met the qualifications. The contract was almost ready to be signed when we found that although the house was wired for electricity, no electricity was available. No way would we settle for a home without electricity. Especially after becoming accustomed to all the nice things it afforded at the house in which we were living.

> **"No way would we settle for a home without electricity."**

    The problem with our future home, we discovered, was the location between Six Mile and Clemson. The power company provided electricity along the highway between the two towns but the house we were considering was about one-half mile from the highway. The distance was too great for the power company to provide a line for running electricity.

    In the late 1930s my husband worked for the Rural Electric Coop, when the lines were constructed in Pickens and Oconee Counties. He helped clear right-of-way, dig holes and

Political Shadows

set poles all over the two counties. He recalls the difficulty of working among the mountainous hills and bluffs around Rocky Bottom. It was his first public employment. He willingly walked five miles from his home near Rices Creek Baptist Church to Easley each morning to catch the work truck and back home when the day's work was done. Quite naturally, he turned to the co-op when the problem of securing electricity arose.

Blue Ridge Cooperative was very cooperative! A line was constructed one mile across forest and field from Maw Bridge road to our future home. The lights came on. The loan was approved, the purchase made.

We moved here December 12, 1952. Mother has passed on and the children grown up and moved on with their lives. Now in our 80s, we are grandparents, and great grandparents several times over. Loss of strength, agility, hearing, vision and memory is taking its toll.

Our lights still glow and with modernized bulbs they shine more brightly that ever.

Wrote Norris about problem

by John Zimbelman
Palmer, Nebraska

Alice Kurz (Zimbelman) and I got married February 7, 1948. We had a hard time.

Her folks lived a half-mile from us. They also didn't have electicity. They had neighbors around that had electricity for years, but they didn't want us to have electricity.

We wrote to George Norris about our problem that we couldn't get electric.

And George Norris helped us.

We got electricity in the Spring of 1950. Our daughter was two months old at the time. We also had a son born in 1954.

Our folks and us were so glad to finally get electricity. We really enjoy our electricity. Don't know what we would do without it.

Door-to-door lobbying

by Eunice Kost
Washburn, North Dakota

I was born on a farm about 15 miles northeast of Washburn, North Dakota, then lived on a farm 11 1/2 miles northeast of here. When I married a farmer, I lived 2 1/2 miles northeast of Washburn and that is where I was living when REA energized the lines to my home Nov. 7, 1947.

What a day. My father Emanuel Kack was one of the pioneers, who worked long and hard to make this available. Many people still doubted it could even be brought to farmers. People kept saying it was unthinkable and impossible. Many people near us doubted it.

My father wanted to bring the line from Washburn, but he said most of our neighbors were rejecting it. They needed every farmer to sign before a line would be set our way. So he asked if I thought I could convince any of my neighbors.

I didn't have the easements to sign, but I went out before them to convince my neighbors we needed everyone to sign. I visited every one of my neighbors to try and convince them to sign the easements to let the electric power lines to be placed on their land. I was a busy farm wife with three small children, 1, 3, and 5. It was not an easy task.

Some said I lived 50 years without electricity and I can live 50 more without it. Others thought it was impossible. Still others had just purchased wind chargers or 32-volt power plants.

I prayed hard as I went from farm to farm. You really had to be a strong believer to outtalk all the neighbors, who were all much older than myself.

Circle of Brightness

REMC made dream come true

by Russell D. Merrell
Kokomo, Indiana

In 1939 I was a young farmer with a wife and one son in rural Howard County, Indiana. We had been married seven years. My wife taught school before our son arrived. Surrounding towns and cities like Young America and Kokomo had been electrified by Public Service as early as 1917. Rural electrification seemed like a dream — something that would never really occur.

On August 15, 1939, the Carroll County Rural Electric Membership Corporation (REMC) made our dreams come true. For years we had tried to get Public Service to extend lines to our area. We made visits and requests, but they required three customers per mile and a $6 per month minimum bill.

We met their conditions but they still refused to bring power to our area.

We were at the west edge of Howard County and some wealthy people convinced Public Service to bring power at the extreme price of $15 per month. REMC saw the need, applied for federal grants, and extended the lines to our area at a minimum cost of $2.50 per month. I spent many hours signing up people for the service. In 1939, people who never thought it would be possible had electric power in their homes.

Carroll County was very progressive in the area of rural electrification. Claude Wickard, the Secretary of Agriculture in

> "We met their conditions but they still refused to bring power to our area."

Washington D.C., was a native of Carroll County and helped to make this a reality. After REMC serviced us, Public Service offered electricity to all of Howard County on the same terms. At a public meeting, I asked what we folks on the west edge of the county should do. The presiding man replied, "I believe I would stay with the boat I'm on." We did and have been with them since.

Before then, local people had lost hope of rural electrification. On our farm, and on many others, Delco® Power Plants, which gave 32-volt electricity, had been installed. When word came of REMC's offer we made plans to trade our 32-volt equipment to a local man for the things we would need to switch to 110 volts.

I spent most of August 15, 1939, changing the motors on our refrigerator and water pump, setting in our electric stove, changing the light bulbs, even converting our 32-volt radio. We installed the meter base, and the man came to put on the meter about 2 p.m. Rural electrification had become a reality.

Of all the changes I've seen in my 90-plus years of life, electricity has made the biggest difference. Farming and taking care of the home became so much easier with electricity. An electric feed grinder was soon installed for my use, and the hot water heater was a joy. In December we even had Christmas tree lights for the first time.

We are grateful to Carroll County REMC for making our dreams a reality and providing us with the life-changing force of electricity.

T.V.A.'s scope misunderstood

by Phyllis Reynolds
Farnam, Nebraska

Many changes were made in 1947 in Farnam, the year REA came to make our lives better. Since REA supplied alternating current, changes with the wiring of the houses, the businesses, and the appliances needed to be made, because the town's power plant generated direct current electricity.

The town's supplier of power was installed in 1913, the same year George W. Norris left his job as a representative and became a senator and began promoting projects such as the Tennessee Valley Authority.

In 1933, his dream was realized but the people of Farnam were not concerned about the TVA. It was hundreds of miles away. They were concerned about their welfare since these were the Depression years. Some of those who kept up with the daily news thought money could be spent to a better advantage elsewhere. They weren't aware of the importance of the project and of the good that would come of it.

They were proud of their power plant that generated 32 volts. It was very important to them because it provided the electricity for their lights and for pumping the water. Their power plant was much more dependable than the windmills they gave up years earlier.

There were two main diesel engines and an auxiliary engine. One engine was operating at all times and when there was a need for more electricity, we started the second engine. The power plant's engineer relied on the gauges to let him know when there was sufficient power to supply the needs of the residents.

Later, when they were made available, the townspeople bought radios, small appliances and refrigerators. The power plant was still able to supply enough electricity for them.

Two days out of the week, the engineer of the power plant knew two engines would be needed and would run long hours. One was Monday, the traditional washday, and the other, Saturday, when farmers and their families came to town in the evening to visit and to purchase farm supplies, and to do their trading in the grocery stores. It really was trading because then eggs and cream were sold in return for groceries and other articles that the general stores had to sell.

But equally important was the theater — the picture show. Every Friday and Saturday night there was a movie shown. People, young and old, would go to the shows whether the movies were westerns, musicals, or war stories.

In 1947, three years after George W. Norris died, REA came to our small town. The war was over and Farnam was getting new businesses. More people were becoming residents in town. Farnam was growing, again.

REA was going to the country too. No more kerosene lamps, cook stoves that needed cobs and wood, and no more Works Projects Administration privies. It was a great time, but it also meant the closing of Farnam's power plant that had been in operation for 34 years.

If George W. Norris were alive today he would be very proud to see all the improvements that he instigated by his involvement in the Tennessee Valley Authority many years ago.

But as I drive by the lot where the old power house once stood, I imagine I hear the steady beat of the big diesel engines still supplying water and electricity to our small town and listening to the residents of Farnam saying, "This is as good as it can get," not realizing REA would make it even better.

Chapter 6

Waiting and Wiring

A new rural landscape welcomed poles and wire.

Circle of BRIGHTNESS

Strung along

With his background on the House Committee on Public Grounds and Buildings, George W. Norris learned of a functioning public power system in Ontario, Canada, in operation since 1908. He saw the project as an example of the limitless possibilities for generating and distributing electricity.

Electrifying the countryside was something other countries were doing much better than the United States, where only 10 percent of rural farms had electric service. In Sweden, almost half of all farms were electrified, in New Zealand, 67 percent, 90 percent in both Germany and Denmark and almost 100 percent in Holland and Switzerland.

By 1922 it was time to do something — especially when he perceived "big business" was moving in. Norris submitted a bill calling for the government to operate federal properties at Muscle Shoals. This was in response to Henry Ford's offer to buy the entire property for $5 million to feed hydroelectricity for private manufacturing use.

In 1925 Norris toured Ontario and not only saw what a public power project could do for a region, he saw rural men and women with pride on their faces and met the people whose monotonous lives were improved with electricity. He saw how liberating it was for them to have their lives of drudgery ended.

It was a long and politically-charged process before the REA would ultimately be established in 1935 — a testament to Norris's pioneering faith.

There was nothing "Roaring" about the 1920s when it came to tangible action resulting in improvements for rural Americans. The 1930 United States Census Report showed the rate of rural electrification was almost the same as it had been a decade earlier.

A rural countryside felt its government had merely been

stringing them along all that time, taunting them with electrical promises.

As the idea of rural electrification moved closer to reality, the Depression broadsided the rural countryside and the idea seemed farther away than it ever had.

Many lost faith. Many felt deceived by their government.

Even when word spread about President Franklin D. Roosevelt's "new deal" and something called the Rural Electrification Administration, it was difficult for farmers to hold out hope for anything other than hollow promises — until trucks brought the first power poles and the first wave of men ready to string the first wire to the farms.

What a Kentucky farmer said about wiring: "We'd heard the government was going to lend us money to get lights, but we didn't believe it until we saw the men putting up the poles."

In George's Words: "I grew up to believe wholly and completely in men and women who lived simply, frugally, and in fine faith."

In the words of the people, here are their stories of "Waiting and Wiring"....

Circle of Brightness

Dad's delayed foresight

by David Parshall
Stanwood, Michigan

In the 1930s, my father owned a water-powered flour mill located on the Shiawassee River in central Michigan. It had been in the family since my great grandfather built it before the Civil War. Though it was outdated, the mill provided us a living during the Depression years, when so many people were out of work and soup lines were common in the cities. We ate plenty of bread and pancakes.

Our home was an old house on the hill above the mill where we lived comfortably but somewhat primitively. My father was an engineer, and he announced that he was tired of the old oil lamps, and would provide us with electric lights.

He designed a small water wheel, and made the forms and had the casing cast at the iron foundry in nearby Chesaning. With the water wheel installed at the mill and connected to a generator, we had electric lights in the mill and in the house!

Unfortunately, it was direct current (D.C.) and the lights were not even as bright as the oil lamps. Also muskrats had a habit of entering the small water wheel and stopping it. Every time that happened the casing had to be completely dismantled. Dad struggled for a long time but his electric system was never a success.

He took quite a ribbing from the family, but could not improve the system sufficiently to make it worthwhile. Dad's ventures into electrical power became a

> **"Dad's ventures into electrical power became a source of family amusement."**

source of family amusement.

    Fortunately, shortly thereafter, it was announced that rural electrification was coming to our area. In anticipation, dad installed a bathroom complete with a sink, shower, and toilet, a kitchen sink with faucets, and a pump to supply water to the house from an outside well. He completed all of this before rural electricity came and the family had great sport asking dad how he liked his shower and did he "remember to flush the toilet?"

    His credibility was low. One astonishing day the power came on. We ran from room to room shouting, "Look at the bright lights! The toilet flushes! The water runs." We all agreed that Dad had done it this time.

    Our house had always been heated with wood stoves, one in the living room and one in the dining room. A cook stove heated the kitchen. There were no fans to distribute the heat. After we had electricity, a coal stove with a fan could heat the whole house. No more hauling wood in and ashes out.

    After the lights came on my grandmother gave us a small refrigerator to replace our ice box. The whole family enjoyed the improvements that electricity brought to us. We had lights, running water, improved heat in the house, a refrigerator, and we could say goodbye to the cold, breezy outhouse in the back yard. What a vast improvement in lifestyle!

    The day the lights came on our quality of life changed from primitive to modern — and Dad was a hero.

War delayed wiring efforts

by Martha Coers
Waldron, Indiana

After the agonies of World War II were over, my husband and I struggled to build a house. Early in 1946 we acquired a piece of land. Finding material was difficult. War-caused shortages were still rampant. We decided to build a small place that could be improved later.

In the spring we began erecting our walls, block-by-block. My husband's father helped him frame the roof. We used salvaged material and begged and borrowed each board and bag of mortar needed for our effort. By autumn we had a building. The outside walls were block; the inside ones were tar paper. Our water came from a pitcher pump, heat came from a coal stove and the kitchen range used bottled gas. Our abode was better than Abraham Lincoln's cabin — but barely.

After the wiring was installed, the inspector came. The work did not pass his inspection because we lacked switch plates. They simply were not available. My husband went to his parents' home, unscrewed their plates, and placed them on our switches. He called the inspector and our electric work was approved.

Then we made our way to the Rural Electric Membership Corporation (REMC) office and applied for a hook-up. Carry-over from the war was evident. The work was quite backed up, much post-war building was in progress, and our name was placed on the bottom of a very long list of hook-up orders.

We moved into our new home on Thanksgiving — thankful to be there, but making do with coal-oil lamps, flashlights, and silence. This pre-dated TV, but we could play no records or radio. I made regular visits to the REMC office, pleading for help.

Then came Christmas. We cut a tree in our woods, decorated it with shining things, but left the light strands in the box. During the day on the 24th, I made preparations, cooking, baking, cleaning and by late afternoon, I was ready to light some red candles.

I heard noises outside and went to the window. I saw a prettier sight than Santa's sleigh — it was a service truck and the power was being connected!

> **"I saw a prettier sight than Santa's sleigh — it was a service truck and the power was being connected!"**

I quickly grabbed the box of light strands and put the bulbs on the tree. When the work was completed outside, I plugged them in. Christmas had come! I played holiday music on the recorder as I thought of the director's command: "Lights, action, music." Now we had them all, just in time for Christmas.

Electrician left souvenirs

by Lorna Graff
Seward, Nebraska

I was born on a farm in 1931 in Nuckolls County Nebraska. At the age of four, my father moved to a farm along the Lancaster and Seward County line. The "dirty thirties" at Ruskin, Nebraska were very hard for my parents, but they always took good care of us.

In the spring of 1939, a man started wiring for electricity in our home. I started saving the little knock-out slugs from the electrical boxes because they looked like nickels. That great night in April 1939, my sister, my mother and I were sitting on the floor . . . waiting . . . when they told us they would turn the lights on. It was such a wonderful feeling when my dad came in and said, "How do you like that?"

> **"I started saving the little knock-out slugs from the electrical boxes because they looked like nickels."**

I still remember my early childhood in the 1930s, on a Lancaster County, Nebraska farm, that had no electricity. I remember my mother filling the kerosene lamps. In one room we had a gas lamp but we had to buy mantles when we used that one, as the gas lamp gave more light than the kerosene lamp. I was a little depressed to have to sit in the dark, so sometimes I sat under the kitchen table playing with my dolls hoping it would not get dark so early.

Winters were cold so we would warm up and dress near the oven of the cook stove.

I recall one night when my father was helping my sister

with her homework. I sat by another kerosene lamp in the kitchen and put a comb across the glass of the lamp, hoping it would get hot to curl my hair like my mom did (she used a curling iron of some kind) but the comb burnt my fingers. I tossed the comb into a corner. No one noticed it at first, but we came close to having a fire! With burnt fingers my dad carried me upstairs to bed while I said I'd never do that again.

By having electricity things seemed so much better. We could flip the switch or pull the string and look out at the yard light. We didn't get a refrigerator or appliances that I remember, because in 1940 my mother died at age 38. We had lots of changes then since my little brother was only 16 months old.

My Daddy and all of us were so grateful that my mother was able to enjoy the wonderful electricity and the pleasures that came with it. Those were very important years in my life and I will always remember "The Day The Lights Came on."

Circle of Brightness

Linemen rarely went hungry

by Warren B. Rogers
Lake Wylie, South Carolina

My family was living in Greenwood, South Carolina when the Rural Electrification Administration came to the county. My dad was one of the first two lineman hired. Contractors were building primary distribution lines throughout the county. Dad's job was to construct "secondary" lines from the primary lines to individual houses.

Most of the time, when the power reached the houses, they had not been wired. Often Dad would make a deal with the owner to return on Saturday to install wiring. As a seven-year-old boy, my job was to go into the attic and pull wire from room to room.

It was not uncommon for us to receive vegetables, chickens, fresh meat or country hams rather than money for the work.

The houses would only have one light in each room. A big spender would also have a light on the porch.

There was no need for receptacles because no one had appliances.

It would be years before many appliances made it into the country, because the country was still working itself out of the depression and the war started a few years later. It was sort of a toss-up between refrigerators and washing machines as to the first appliance.

> "It was not uncommon for us to receive vegetables, chickens, fresh meat or country hams rather than money for the work."

Another task of dad's was to read the meters each month. He was also expected to collect the bill on the spot. Most bills were 75 cents — the minimum.

It was a mystery to Dad why, after a lifetime of no electric power, many people would call in the middle of night to inform him that the power was off. The remarkable thing is that people would go to major trouble to get to telephones since there were few phones in the county.

Another perplexing situation was several men throughout the county complained the electric lights were bothering their chickens, and egg production was down. It seemed strange because there were no lights in the hen house and we were sure the chickens did not live in the main house. Disconnecting the electric power was not an option the men liked.

The most frequent change I remember noticing in the houses as we turned on the lights, was the lamp table in the center of the parlor was moved somewhere else. This often permitted a major re-arrangement of the room. People started staying up later since they had better lighting by which to read or work.

Circle of Brightness

Mother kept linemen on task

by Joy Siekirk
Comstock Park, Michigan

In the summer of 1948, I was 15 years old. We lived on a small farm one mile east of the town of Irons. Our farm was the last one on that line to get electricity. Our house was wired by my uncle several months earlier. The electric poles were already in place along our country road. We were waiting for the linemen to install the electric wire, then we would have electricity.

My mother was tired of seeing her new washing machine standing in the corner, and not able to use it. She could hardly wait to flip a switch and have the lights come on. My young thoughts were to have a nice cold dish of Jello® in the middle of summer.

That day, the linemen were putting up the lines, so we expected to have the lights that night.

My mother sent me to the store to get a few groceries and a package of Jello®. I decided to walk the mile to the store instead of riding my bike.

As luck would have it, our pet deer decided he wanted to go for a walk with me. He followed me all the way to the store and back again. I felt like Mary and her little lamb. I could hear the men laughing as I walked by. It was nearly 5 o'clock, they had all the wire up and were at the last pole by our house when I got home. The men got down from the poles and started petting and playing with the deer.

> "The men got down from the poles and started petting and playing with the deer."

I can still hear my Mother yelling at me to get that deer in the back yard. She thought the men would quit work for the day and we wouldn't have our electricity again that night. The lineman said, "Madam, you have your electricity, all I have to do is put the fuse in your fuse box."

We had lights that night and Jello® the next day.

Circle of Brightness

Electricity changed railroads

by R.J. Heinle
Cadott, Wisconsin

On October 20, 1936, at the age of 15, I went to work as a lineman on the DSS & A Railroad at Marquette Michigan. This was a branch of the Soo Line Railroad.

On Dec. 7, 1941, I was foreman on a crew rebuilding the line from Abbotsford, Wisconsin to Ashland, Wisconsin.

Then, I enlisted in the Marines the following week, leaving the crew a few weeks later at Phillips, Wisconsin. I returned from the Pacific and was discharged November 25, 1945.

Instead of returning to the line crew, I transferred to the signal department. They also took care of electrical work.

The first of January in 1948, I was given a bunk car and three box cars loaded with electrical wiring materials. I had orders to wire all the depots, round houses and bunk houses on what was known as the "wheat line." This started at Thief River Falls, Minnesota and ended at Kensmore, North Dakota.

The depots all had living quarters upstairs for the agents and their families.

The REA ran line to the rural area at that time, so the rush was on to light them up.

All lighting was kerosene wicks or mantle lamps, a few Coleman® lanterns and lots of candles.

I was told to just put in the bare necessities, which were two outlets in the office, a center and desk light, all cord and pull chain.

While wiring the living quarters, I was swamped with cookies, cake, coffee, and a lot of great meals, as the agents wives were hoping for a few extra lights and outlets. I never disappointed them.

Waiting and Wiring

> "When I would turn the lights on for the first time, there was always an audience...."

When I would turn the lights on for the first time, there was always an audience, section crews, local people and of course the agent and his wife. She generally broke into tears.

It was a rewarding experience to make that many peoples' lives more enjoyable and happy.

North Dakota is noted for its cold temperatures. February 1 of 1948 was no different, staying between 20 to 40-below zero. I moved to Fordville, North Dakota to wire a 10-stall roundhouse. Soot hung from every timber, as it was close to 70 years old. The roundhouse crew could not have been more cooperative. When the wind blew the snow through the walls of my bunk car, they pushed it into the roundhouse, very dark and smoky but a lot warmer.

The roundhouse foreman was Archy Helgerson, who worked there for 40 years. At his retirement party in about 1956, he was asked what his greatest experience was in all the years on the railroad. He responded "The day they turned the lights on." To me that defines what the REA meant to the millions of rural residents.

I retired as signal mainter in Chippewa Falls, Wisconsin in November of 1980. — just in time for deer season at my camp in Phillips, Wisconsin. I've been an REA member since 1949.

Circle of Brightness

Using rats to string wire?

by Eleanor Zywic
Columbus, Nebraska

We were waiting. . . waiting. . . waiting. . . so we could turn on the lights in our house, and live like our relatives and friends in the city.

Oh happy days, when those electric guys arrived with the tall poles and wires right up to our house, ready to saw the holes in the walls and to put through the wires. One of those guys asked us to quick, catch a rat, so he could tie the wires to its tail and pull the wiring. And we believed him. Next thing, the wires are pulled in (without the rat) connected to the switches and . . . Wah-la . . . magic lights everywhere, indoors and outdoors, the whole yard can now be lit up!

Now the whole house could be lit up too— not just that little old kerosene lamp in the room we were in. No more groping in the dark to get to our beds at night. Goodbye to that dirty job of filling the lamps with kerosene and washing sooty lamp chimneys — gone forever. Imagine not only bright lights, but no more oily smell in our house, and now we could go buy all those plug-in appliances and things. Gee! At the flip of a switch, hot and cold water at our fingertips, food cooking, and heating the house without putting those cobs and wood logs in the stove. Whoopee — no messy old ash pans to carry out!.

And the greatest of all, no more chamber pots to carry in and out of our bedrooms or running way down the path to the outhouse sitting on that cold old seat. Ooh. How nice to sit on a comfortable seat and flush. No more being afraid to go outdoors after dark. Now we could slowly come back into the house without running ever so fast, and wondering if we could squeeze through the door fast enough.

"And let there be light. And light there was."

Skinny cousin useful

by Stella C. Moore
Tucker, Georgia

I was born in 1926, seven miles out in the country from Jasper, Georgia. By 1940, we still had no electricity, only kerosene lamps and a refrigerator, which had to be filled every two to three days, and the lamp wicks trimmed.

Amicalola Electric Membership Corporation ran lines into our community. Our house was not wired, but an older cousin knew how to wire the house came and I helped him. I could crawl in tight places and pull the wires and return them where needed.

> "I could crawl in tight places and pull the wires and return them where needed."

When all was finished, it was so great to have electricity — just pull a cord or switch, and have lights and fixtures to see by and use.

It is great how God has given man wisdom to do so many wonderful things.

Circle of Brightness

Teen built, wired new home

by Oral Rattiff
Lizton, Indiana

Fall 1936.

My father saw in the county paper that Rural Electric Membership Corporation (REMC) were putting power lines through the county.

Dad said we would just tear the house down and build a new one.

REMC taught school on wiring one night a week so my cousin and I signed up.

We moved into the garage while we tore the house down. My dad and I cut logs for floor joists. The sawmill cut 16-foot-long 2-by-8s of white oak.

Dad, the neighbors, and I dug the basement with two teams of horses, two slip scoops, and a breaking plow. We got the wall up. It was a heavy load of lumber for a 16-year-old, 140-pound boy to handle by himself.

We put the floor joist down and I crawled out to the end of each one and bored a one-inch hole using a brace and bit.

> "It was a heavy load of lumber for a 16-year-old, 140-pound boy to handle by himself."

We hired a neighbor and four others for 50 cents an hour to frame the the house. When it was done, the neighbor fired all his carpenters because they were not neat enough to do finish work but his work was beautiful.

Dad gave us $25 for wiring. We made a ladle to solder the wires and we used a blow torch to heat the solder. Dad told me take the truck to the creek which had washed gravel. I took

Waiting and Wiring

a 2-by-12 so I could back out on gravel. I scooped 1.5 yards each truck load. Using muscles, we cemented the basement floor.

The house was 28 by 46. Next, we plastered the walls. and had to go to Danville to get an additional $25 worth of wiring and light fixtures.

We moved in 1936. Original cost of the house was $1,500.

Dad came to pick me up every evening after basketball. Every evening. I asked if they had turned the power on yet and every evening he said, "No.' This went on for several weeks. Quite a bit later I asked him, "Are they ever going to put them on" and he said, "No."

So that very night as we came home my parents had every light in the whole house burned.

Wow. We did a good job wiring!

Later on my cousin and I wired a few houses, but had to quit because my cousin left to attend Purdue.

Some years later, my wife and I moved into the house and did only two things to — we rearranged the kitchen and rearranged the bathroom.

Circle of Brightness

Linemen brought 'magic'

by Gertrude Clay
Lugoff, South Carolina

    Yes, I remember when the lights came on our road in Russellville, South Carolina, in Berkley County. I was nine or 10 in 1941, when I saw men on our streets putting up poles and wire. I couldn't imagine in my mind that we could have light at the push of a button. We had kerosene lamps that had to be filled everyday, and the globes had to be washed in vinegar and soap to get the soot removed.

    They came into our house and wired one electric cord in each room, dropped from the ceiling with one light bulb and electrical socket. When they finished, I immediately pulled the string hanging from it to turn it on. At that time it was not working, but when it did work I thought it was like magic. Everyone had to pull that cord two or three times a day to see if the "magic" still worked.

    We didn't have anything to plug into the electrical socket. We kept things cool by buying a block of ice and keeping it in a tin tub, covered with a blanket. In months to come we were fortunate to get a refrigerator and radio. They plugged into the same socket that the light was on. Daddy bought a double socket that screwed into the light bulb. This would accommodate these appliances and burn the light at the same time.

    We were the only family around that had a radio (we used to run it off a car battery). People would come from miles around to listen to the "Opry" on Friday Nights and we would listen to the news. This very radio was the one we learned of World War II. I remember thinking in my child's mind that one day we would see what they were doing on that little dial. After electricity came, our radio socializing came to an end, as

Waiting and Wiring

> "Medicine shows also used our electricity in exchange for elixirs and tonics."

everyone seemed to have a radio then.

There were tent movies that came to town showing all westerns. They would give us passes to the movies for that week if we let them plug into our electricity. Medicine shows also used our electricity in exchange for elixirs and tonics.

Of course it was several years before we had indoor plumbing for kitchens and bathrooms. My momma finally got an electric iron and we all felt that was a great invention. Before that we heated our flatiron on the kitchen wood stove. Later, we had all the appliances and things that are taken for granted now. We no longer had to hand pump gas, we had freezers for meats, jukeboxes for dancing and fans for cooling. Oh it was just a wonderful day when the lights came on. We are still enjoying our electrical delights. We had all of this because of the REA.

Circle of Brightness

Life-changing energy

by Elmer Saltz
Norfolk, Nebraska

It is vivid in my mind the image of the electric lines closing in on the farmstead where I lived. It was 1948 and I was eleven. I can still visualize the poles and the lines being set in place.

My folks hired a contractor to wire the farm. In those days this was not an easy task; most of the wire had to be threaded between the wall studs and laid in the attic on top of the ceiling joists. The house was being wired when the electric lines were only a few hundred feet from the house. The contractor had a small engine connected to a generator. We persuaded him to let us use it for a short time that evening before REA finished the task of running lines to the farm.

We could only use two light bulbs, So we used them in the kitchen and the living room. How bright the rooms seemed!

We even noticed how grimy the walls and the ceiling were. This was common, of course, in homes where the source of heat was a wood and coal-burning stove which was often used to burn cobs picked up in the hog yard.

A refrigerator became the first electric appliance we purchased. That ended us going to town and buying big blocks of ice. At this time I did not know that Hans Christian Oersted, a Danish physicist, had shown in 1820 the relationship between an electric current and magnetism. Many scientists at that time started working on developing a theory of electrical phenomenon and others worked on finding methods of developing this technology.

A washing machine was the next household purchase. What a great improvement over the stubborn Maytag® which used a one-cylinder engine to run the old washing machine. I should not forget the yard light. One could now go out after dark and see. Now the farmstead didn't seem nearly so scary after dark.

Next came an electric cream separator. It seemed so gratifying to turn on the switch, let the separator bowl come up to speed and turn on the milk spigot and stand back to watch the cream being separated from the milk. Most of the cream we sold in town to meet living expenses.

An electric milking machine was purchased some time later. How great it was! One time when my dad oiled the pump he did not replace the belt shield. Some time later, I was placing the milking machine on the cow stationed adjacent to the pump. I looked up when I heard a thump and saw the end of her tail flying across the barn and striking the wall. She had swished her tail and it became entangled in the milking machines drive belt.

A year later, we moved to another farm that had a water pump in the basement and an indoor privy. I have not forgotten about gong to a well some distance from the house and bringing water to the house, since we had no electric water pump. Nor have I forgotten going out to the outdoor privy.

Over the years, I find myself taking electricity for granted, until there is an electrical interruption and that drives home the importance of electricity in our lives.

Today I would miss heat in my home, television, satellite reception, warm tractor radiators, heated stock tanks in the winter and the list can go on and on. Once nature yielded its secrets about electricity, then man had to learn to harness water, wind, coal, and gas energy and later the energy found in the atom had to be converted to electricity and distributed to the consumer. I am thankful for the men and women that had the vision to develop the electrical systems we have today.

I am grateful for their wisdom and foresight.

Circle of Brightness

Oklahoma linemen remember

by Cathey R. Heddlesten
Wilburton, Oklahoma

Two retired electric cooperative linemen — who helped bring power to rural consumers in the late 1940s — have fond memories of "The Day the Lights Came On" in southeastern Oklahoma.

"It gave me the biggest thrill to watch the elderly consumers' reactions when the first electric lights were turned on in their homes," said B.T. Davis of Howe, Oklahoma.

Davis began his career in 1948 with Kiamichi Electric Cooperative of Wilburton, Oklahoma, when the system was in its formative years. He worked 41 years as a lineman and area maintenance man before retiring in 1989.

> **"It gave me the biggest thrill to watch the elderly consumers' reactions when the first electric lights were turned on in their homes."**

"I remember, in particular, one elderly woman who grabbed her husband around the neck, squeezed him hard and said, 'Grandpa, I didn't think we'd ever see light like this in our house,'" Davis recalled.

Davis said many residents were initially frightened at the prospect of allowing electricity into their homes.

"We'd have to go in and show them, by pulling the light chain, that it wouldn't hurt them," Davis said. "They were afraid it (electricity) would kill them and they were afraid to pull the chain."

Davis said the biggest effect electrification had on the

lives of rural Oklahoma residents was the addition of luxuries like lights and iceboxes into their homes.

"People were so very grateful to finally be getting what the townsfolk had gotten as early as the 1930s," he said. "The old-timers would tell us not to bring a lunch on the day we turned on electricity to their homes because they wanted to feed us — and they always fed us like kings."

James White, who started work for KEC in 1949 and retired in 1995 after working the majority of his career as an area maintenance man, said he also remembers the joy expressed by area residents when they received electric power in their homes.

"The people were really happy," he said, "even though most of them had nothing but a radio and maybe a bare light bulb or two at first."

In his family home, the first two purchases his parents made after their house was energized were an electric iron and a radio.

"A refrigerator came later," he recalled. "Back then, the minimum bill was $3 for 25 hours or less of electricity. People stayed with items that didn't pull a lot of electricity."

White says the biggest benefit of rural electrification for the majority of southeastern Oklahoma residents was the improved quality of lighting in their homes.

"I heard lots of people say how much better it was reading by that bright electric bulb than it was by coal oil lamps or candles," he said.

White said the electrical power that is often taken for granted today was very much appreciated by southeastern Oklahomans back in the late 1940s.

"Younger people don't know what it was like to live without electricity," he said.

"They're only concerned now when the lights go off."

Circle of Brightness

Bookkeeper kept long hours

by Virginia Guretsky
West Point, Nebraska

I was the first bookkeeper and stenographer for the Cuming County Rural Public Power District, as it was known when I began work on April 15, 1938.

I was approved by the Rural Electrification Administration in Washington, D.C. to fill the position and worked long hours. I worked from 8 a.m.-6 p.m. six days a week and worked until 10 p.m. until the 15th of the month when a penalty was added for the unpaid bills. I also worked many evenings to get all the work down.

I set up the double entry bookkeeping system for the CCRPPD and also set up all the accounts for the customers. I typed right-of-way easements and checked their correct descriptions.

Due to my experience in the law and abstract business with my father, Charles Beckenhauer, I knew how important it was and what a problem it would cause if these descriptions were incorrectly recorded at the courthouse on the wrong land.

I helped distribute payment of all bills, salaries, and director's fees and all correspondence. All this I did on a manual typewriter, not an electric typewriter as it is done today. I also helped line up the program for the turning on the lights on Aug 30, 1938 during the Cuming County Fair. We had 238 customers at the time. I addressed meter cards and bills, all done by hand at the time.

As additional customers were added to the present lines, I worked with the linemen making work orders and drawing diagrams and all the paperwork involved in adding new customers.

Shortly after beginning operation September 1, 1938, we

Waiting and Wiring

put out a Cuming County Rural Public Power District paper each month consisting of four to six pages at the time. I wrote many of the articles and typed all the stencils. These were mimeographed on a hand turned machine, not on a computer as used now. These were addressed and sent to all customers.

A second girl was hired about six months after I started and she worked under my supervision. I began work April 15, 1938 and resigned in April of 1942.

> **"Even though I resigned in 1942, I found myself promoting rural electrification even 50 years later."**

Even though I resigned in 1942, I found myself promoting rural electrification even 50 years later.

I even helped line up the 60th anniversary of Cuming County Power District, held in the towns of Cuming County.

I attended all the celebrations and visited with people who were there. I was the only person living from the time the Cuming County Rural Public Power District started.

From assistant to technician

by Ted Shonts
Colorado Springs, Colorado

I was 15 when my dad was notified that we would be getting electric power to our ranch. This was great news for me — it meant that I might no longer be required to pump water by hand on some of those cold winter days when there was not enough wind to turn the wheel on the windmill. It meant that I would no longer need to fill the kerosene lantern with oil and do all my chores by that small light, to milk cows in that dim glow. Needless to say, I was excited about this new development.

The first consideration was to get the house wired for this new utility. The job fell to my brother-in-law, Carl Wertz who nominated me to be his assistant.

At that time I knew almost nothing about electricity except that Model-T Ford coils made good shockers.

Our home had originally been a homestead which had been built with huge blocks cut from sandrock found on the property. Obviously, there was little choice on where to run the wire, so we cut channels in the wall, installed the wires and plastered over the channel. As I crawled around the spaces in order to run electrical wires, I discovered many crooks and crannies in that old home I never knew existed. I even discovered some charred 2-by-4s in the attic where there had been a near disastrous fire years before my

> "At that time, I knew almost nothing about electricity except that Model-T Ford coils made good shockers."

time.

    I learned much from that experience. Enough to help me get into electronics school in the Army where I became a missile electronics technician. Since then, I have assisted in wiring a number of homes including my own.

    The day finally came when we turned on the first switch and stood in awe at the great amount of light we had compared to the old kerosene lamps. I hated filling those lamps especially when the five gallon can was full. It became so heavy I was sure to spill some of the oil and catch "what for" for doing so. This new power also made it so I could have my own radio in my bedroom, rather than crowd around the old battery operated set in the family room. It also meant that dad would no longer have to spend his hard-earned money on batteries for that old set.

    To this very day, I seldom turn on a light switch without a silent thanks for those folks who made all this possible.

    I even feel a little sad for those who never lived without electricity. If they did, they would be more grateful.

    My new home, which is just a mile from the old homestead, has electric cooking, electric water pump and yes, even a computer run by electricity provided by our own Co-op.

Chapter 7

Let There Be Light!

Norris with his ally FDR (second from right).

Norris with McCook friends Carl (left) and Fred Marsh.

And light there was

A decade after George Norris first started his push for the Tennessee Valley Authority, he found an ally in New York Governor Franklin D. Roosevelt, and they shared their hopes for Progressive reform including the public development of the Tennessee Valley for flood control and for harnessing electrical energy.

Two years after Roosevelt was elected President, his New Deal incorporated some of Norris's hopes. Just 69 days into his "First Hundred Days," Roosevelt signed the Tennessee Valley Authority Act, creating an opening for promising rural electrification possibilities.

Roosevelt's right-hand man was Morris Cooke, who convinced Roosevelt that as part of a public program, rural electrification could be achieved to a level of up to 75 percent during his term in office. Farm groups joined the effort.

The TVA demonstrated that a publicly-developed rural electrification cooperative could work when, in 1934, it successfully tested the experimental Alcorn County Electric Cooperative in Mississippi.

That experiment uncovered a concern about rural residents being able to afford the initial investment and cost to buy new electrical appliances and farm equipment. As a result, subsequent efforts would need to address low-cost financing and long-term loans to the cooperatives with minimum interest.

Roosevelt created the executive order in 1935 establishing the Rural Electrification Administration. A year later, Norris and Texas senator Sam Rayburn authored the Rural Electrification Act setting aside $100,000,000 to light a nation's countryside.

When they finally received electricity, it was the closest thing to a miracle many rural residents ever experienced. Some likened it to the miracles they read about in their Bibles, and from the Bible came an appropriate response to their modern-day miracle.

"Let There Be Light!"

What Sen. Sam Rayburn said to Norris when a conference committee on Rayburn's House Committee of Interstate and Foreign Commerce stalled: "Now Senator, don't be discouraged . . . we will come together because we have made up our minds you are not going to give up."

In George Norris's Words: "My eyes had been covered with bandages and I was in complete darkness; but I realized that (my oldest daughter) Hazel was standing by my bed and I said to her:

'Ought you not go out and play with the other little children — those guests who have come to play with you?'

She very solemnly and calmly replied she would rather sit in the room with me. I still tried to persuade her to join the other children, and finally she said:

'Father, if you want me to, I will go out and play with them, but I would rather stay here with you.'

Through darkness I saw light that I had not seen before."

In the words of the people, these are their stories of **"Let There Be Light!"** . . . .

Circle of Brightness

The day the darkness left

by Dorothy Urbanec
Pender, Nebraska

I was born and raised on the farm where my father had lived his entire life. Darkness was a way of life. An outhouse was a necessity. While I was a junior in high school in 1941, the R.E.A. poles and lines were erected. What excitement! My father helped put up poles, bringing a few extra dollars. After having come through "the dirty 30s," this was a thrilling experience of anticipating the promise of unlimited light.

Each farmer chose as to whether or not he wanted to become a customer. Some farmers chose not to be hooked up but since my father was always ready to enjoy a new lifestyle, he elected to receive the service. Having just come through the Depression, the commitment of $3.50 a month sounded like a great deal of money to spend on something we had always done without. Three dollars and fifty cents was the minimum fee and we were sure we would never use more than that amount.

Anyone who knew anything about wiring a building was employed to get the power into a home. Never having had this convenience, we had no appliances so there was little thought of putting in outlets.

After the house was wired and a yard light was put up, the waiting began. Not having had many comforts, the anticipation of light at the flick of a switch was extremely exciting. Being very conservative, most farmers equipped each room with a simple fixture with a bare light bulb in the middle of the ceiling, most with only a pull cord. We had limited access to materials because of World War II.

Finally the big day arrived! I came home from school and there was electricity in our house! We were warned not to

Let There Be Light!

waste the precious commodity. Everyone in the family felt we were very privileged; we no longer had to clean the chimneys or fill the oil lamps. We weren't intimidated about going out to the "eerie outhouse." We could now flick on the yard light and see our way.

Since there was little money and we were accustomed to doing without, there was little thought of planning for appliances. A radio was a luxury and an appliance we all enjoyed. As I remember, the first major appliance my parents bought was a refrigerator. After all the years of storing food in the cave, the convenience of not having to make trips into the darkness to bring food for preparing a meal was a real comfort.

All this happened in the spring of 1941. The next year we entered the war when no appliances were available. All energy and money was spent on the war effort with little thought of making our lives more comfortable.

After the war, the economy improved and new appliances were designed — making life on the farm easier. The darkness had changed to twenty-four hours of light and farm families were afforded the lifestyles of their city relatives.

It was a gradual change but farmers built bathroom facilities, got electric ranges and electric engines on their washing machines.

Today, the agricultural population enjoys the same modern conveniences as anyone. Televisions, computers, microwaves, automatic washers and dryers are found in most farm homes. Outside, we have grain dryers, automatic waterers, pressure pumps to supply water to all the buildings and irrigations for the crops.

Since 1941, we have moved from the darkness into the light. Electricity has made life on the farm easier and more comfortable. We are now blessed with all the luxuries we choose to enjoy.

Circle of Brightness

One-cord, 40-watt luxury

by Mary Kennedy Lewis
Rock Hill, South Carolina

    Joys of childhood flood my memory as I recall sitting around Granny's kitchen table. The light of her one and only mantle lamp was absorbed by each of us. Papa read the newspaper aloud as granny served or mended, my young aunt and uncle prepared their high school homework and I watched.

    In 1938, electricity lines were strung along our highway and Granny and PaPa had the luxury of one cord embellished by a 40-watt bulb hanging from the kitchen ceiling. Over time other outlets were installed and before the war my aunt gave Granny an electric refrigerator! Such a modern miracle was placed in the dining room to protect it from the heat generated by the wood range, on which three meals a day were cooked.

    Before that time, kerosene lanterns provided light for morning and evening chores as cows were milked and animals fed. Wick lamps lighted the kitchen as breakfast was prepared. Dressing was done in the dark as the cold winter progressed.

    A few years after us, electricity came to my great uncle's home which was some distance of miles from the main highway. Some in the community could not afford the expense even then.

    It was after the war that my grandparents got indoor plumbing. Parts and electric pumps for the well were not available from 1941-1945.

    Returning from the South Pacific, my uncle installed Granny's first indoor plumbing in 1946 — which PaPa refused to use.

Light flooded dark places

by Mary Alice Bridinger
Morley, Michigan

Grandma Keech was my childhood care giver and her farm house was always warm in the morning from the pancakes prepared on the keen wood stove. The dining room was also heated with wood, but the stove was fancy and the rim around it was used for drying wet socks when the men came from the barn. On special days we used the parlor's extremely fancy stove to heat the room as we entertained company.

All the rooms were lit with kerosene lamps which sat on an end table, dining room table or on shelves on the wall. Lanterns were carried to the barn for light as well as along the path to the outhouse. Uncle Herbie carryied it by a swinging handle called a bale.

> **". . . I cried myself to sleep because the harsh lights had destroyed my soft and cozy refuge."**

The day the lights were turned on was exciting but that night, I cried myself to sleep because the harsh lights had destroyed my soft and cozy refuge.

Circle of Brightness

Even the shadows changed

by Gay Price
Midland, Texas

I was a school child when electricity came to our farm in Dodson, Texas. Until then, we had kerosene lamps which cast giant shadows on the walls and bathed the rooms with yellow light.

The shadows were often scary, large and flickering, beyond the pool of light that never quite reached the corners of a room. My father taught me to make "hand pictures" on the wall so I wouldn't be afraid of the shadows — a coyote, a chicken, a bird.

My worst experience with kerosene, however, came on a Sunday morning after my mother had dressed me for church. "You may go outside and play," she instructed, "but you cannot play in the water." Unhappy with this restriction (since my favorite pastime was making mud pies) I contemplated how I might stay true to my mother's instructions and yet enjoy my preferred play activity. Water was out. She had been most definite about that. But I had watched her fill the kerosene lamps from a barrel with a tap. The liquid looked about the same as water. With this faulty reasoning, I prepared my dirt and held it beneath the barrel tap. Unfortunately, the barrel was at some distance above my head, so that when I finally succeeded in releasing the stream of "substitute water" it cascaded into my pail, over my arms and down the rest of my body. This was okay with me at first, since I frequently doused myself in water. However, in short time, my skin began to burn, and I felt it prudent to report to my mother. Horrified, she stripped me and dunked me in soap and water, scrubbing vigorously (which of course failed to soothe my burning skin.)

It was difficult to read or play board games at night. One

had to put the book or game close to the lamp and its circle of brightness which was still considerably dimmer than one might wish. My father and mother worked until dark, so our "fun times" were always at night. Only then did we study or do leisure activities.

Electricity changed our lives in the most astonishing ways. It extended that leisure time. We were captivated by our nights turned to days.

Most importantly, the shadows changed. Instead of a golden glow that failed to dissipate the shadows, the rooms were filled with a bright, white light. Never mind that the white light did not compare to today's standards; it seemed to fill our lives with brightness. My books were bathed in that light, and I became an inveterate reader. We purchased a radio and I listened to my two favorite radio shows: "Let's Pretend" and "Inner Sanctum."

But the best thing in my childhood memory which electricity brought to us was the welcoming shining from the windows of our home. At night, when one of us returned, we saw that home from a distance. . . the light shining like a beacon, calling us into that warm, well-lighted place.

Sister's magic command

by Isabel Wright
Cass City, Michigan

I remember when my sister said, "Let there be light!" and there was light.

Growing up as children on the farm, we gathered around the old oval kitchen table each evening. Here is where our homework was done, as well as sewing, reading and writing, with light provided by smelly, smoky kerosene lamps.

Naturally, we were all excited when they finished wiring the house and barn for electricity. It was a fall evening and darkness had set it. We were all gathered around the kitchen table as usual. Our parents had let one older sister in on the secret.

She suddenly jumped from her chair and began yelling, "Let there be light!"

She ran from room to room flipping switches. The dim kerosene lamp light was nothing compared to the light generated from one electrical ceiling fixture.

> "She jumped from her chair and began yelling, 'Let there be light.'"

We checked the yard light located between the house and the barn. It was unbelievable! It actually seemed to be daylight in the circumference of one light's bright rays.

This memory came to mind as I was writing, by kerosene light, when the electrical power was interrupted by a sudden heavy snowfall.

The length of time electricity was off seemed like an eternity. I found myself saying, "Let there be light!" Because of the prompt repair crews, once again there was light!

The day the lights came on

by Karen Hawkins
Leesville, South Carolina

I remember when there were no lights,
    I remember the kerosene lamps shining bright;
I remember wood-burning stoves in the days of old,
    the winter night's floors with holes, oh how cold.
Then as I grew older, the lights came and it was neat,
    it gave us a way to cool and to heat.
Thanks to all who have made it possible to provide
    to a growing community who takes it all in stride.
When our lights came on it was good to see,
    that looking back,
        some things are still hard to believe!

Circle of Brightness

A different glow to our life

by Carl E. Person
Rapid River, Michigan

Our three-room log home, without electricity in 1931 wasn't all dark, but our primary lighting was either the brilliant glare of a hissing lantern burning white unleaded gasoline, or the flickering illumination from the wick of a kerosene lamp.

Reading with the aid of the kerosene lamp was preferred over gasoline because it gave a more intimate light and wasn't unbearably hot to be near as was the gas; however, the uneven kerosene light often resulted in headaches.

Our outhouse was at the far end of the barn, and in the winter, with the lantern, the long trip to the outhouse yielded an unending supply of scary shadows against the white snow. None of us children wanted to make the trip alone.

First, poles were set, and then an electrical power line was run to our house and everything changed. It seems primitive by today's standards, but we used a 60 watt light in the middle of the ceiling of each room. We never advanced to larger sized bulbs in that house, probably because in our Depression-era mentality that would have been deemed more light than was necessary.

After wall plugs were installed, we added floor lamps, which were a great comfort. Then we acquired a refrigerator with a freezing compartment of about ten gallons. This meant

> "... with the lantern, the long trip to the outhouse yielded an undending supply of scary shadows against the white snow."

Let There Be Light!

we could store ice cream, which was impossible with the old ice box. Also, now an outside pole light illuminated the path to the garage, the barn, the wood shed and the outhouse.

After we wired the barn and hen house, the chickens laid more eggs because the lights gave the hens the perception of a longer winter day at our far north latitude. Radio became our preferred evening entertainment. One could be busy with other things while listening to the family's favorite radio stations from Chicago or New Orleans, which came in as clear channel AM stations after dark.

Then a phone line was brought in, strung on the poles that carried the electric power. Changes seemed to be coming very fast, and although this was at the beginning of the Great Depression all of these changes were for the good.

As a student, I appreciated that reading at home had become pain free. Things were never again the same after the day the lights came on at our house.

Circle of Brightness

Saw things in a different light

by Connie Holden
Kershaw, South Carolina

    Our house was so dark, we had to take a lighted kerosene lamp into every room we went into after dark. The kerosene lamp wasn't very bright and finally we got an Aladdin® lamp, which helped.

    Our community was much happier when we got electricity. Church service was much better than with the Aladdin® lamps we had. The preacher even looked better when we got electricity in 1938.

> **"The preacher even looked better when we got electricity in 1938."**

    It seemed so good for lights to come on with the flip of a switch, and not have to fill up a lamp with kerosene. We have so many conveniences now. We don't even have to cook on an old wood stove and an electric refrigerator is just great.

    We have such a good group of workers that keep everything working. They are always cheerful and happy and seem to enjoy their work. We almost are never without electricity. The workers are always happy and seem to enjoy their job.

Knowing 'pitch dark'

by Pearl A. Simmons
Pickens, South Carolina

I have thought about our electricity many times, and how thankful I am for it. It was pitch dark before we got it. I wonder sometimes how we made it.

My dad and I would go out in the pasture and take a toe sack and hunt pine knots to put in the fire. They made more light than the lamp did.

I thank God for that lamp too. I couldn't hardly find the bed sometimes it was so dark. I was 14 and living in the Blue Ridge Community when electricity came on. That first night and ever since, I sure have been thankful for it. I stayed up almost all night, because I could see where I was. I'd go out on the porch and then come back inside. I just couldn't believe it. I could see things I didn't know we had.

> **"I couldn't hardly find the bed sometimes it was so dark."**

All we had were lights for a while because we didn't have anything to plug in. Soon, we bought a small refrigerator, and we could see and eat ice. That sure was heaven on Earth, and I've been enjoying it ever since. Just look what we can do with it, and what happens when we're without it.

Lights offered security

by Dorothy Carmann
Riverdale, Nebraska

Yes, I remember years of farming without electricity, and how much joy it is to live with electric power thanks to George W. Norris, the legislative father of rural electrification.

There was an incident during high school when a yard light would have been nice. I was late milking for some reason and recall walking between a horse and a mud puddle. Father came to carry me to the house. I hated to miss school that Friday but my headache was terrific — I still have the scar on my chin.

I had two ambitions: to teach rural school and to marry a farmer. Rudolf Carmann and I were married in May 1936 and my father rented us this homestead place that is still my home. The barn burned by lightning in 1930.

All of those early years were a struggle: doing chores before dark, using lanterns, washing lamp chimneys, keeping a supply of kerosene on hand and matches. In summer, we had to get ice blocks for the refrigerator and freeze ice from tanks and hog waterers in winter.

Rudolf fixed up a wind charger on top of a building by the windmill, to charge big batteries to run our radio and a pull cord light inside our kitchen door. We could listen to the "Lutheran Hour" on Sunday afternoons and "Lux Radio Theatre" and other programs in the evenings.

Rudy was a good steward of the land, and cooperated fully with all the fellows planning location of the poles in the yard — one of which, Milford Gulleen, would have another important impact on our lives. As they wired the buildings, they dealt with dust in the old house ceilings and walls.

Finally came the day the lights went on, and we could

celebrate with grateful hearts. I made a chocolate cake, and had corn bread, and navy beans and ham for supper.

Rudy bought a refrigerator before the week ended

There was no end of comfort and joy to be able to plug into electric power and have it spread to every building on the place. Rudy changed each one through the years, save for an old granary that the tornado demolished in 1990.

> **"All of the farm programs will never match the blessings of electrical power on the farm."**

All of the farm programs will never match the blessings of electric power on the farm.

The man who helped set poles for us in 1949, Milford Gulleen, had a daughter, Marcia, who grew up to marry my youngest of six children, Larry.

Circle of Brightness

Out of darkness

by Carolyn Irene Barr
Hemingway, South Carolina

    Before we got electricity, we had to get up very early and do things while it was light, including our homework. And then when it became dark, we stayed up for a while and went to bed.

    After we got lights, things changed in our community, everyone was happy, we were no longer in total darkness. We received our electricity in 1954.

    That's how it changed for me.

Let There Be Light!

Daylight extended

by Juanita Davis
Pamplico, South Carolina

During the early 1950s, we did not have any electricity. We used kerosene lamps and lanterns for light. We had to take the lamps from room to room, because we only had two. The house was pitch dark until the moon was shining. We had to do all our house work — even cook before night.

In 1955, we received electricity in our home and community. In those days, the houses were about a mile apart or farther. The day we received our lights, I will never forget.

It was like everything was so bright to me.

It was like a blind person seeing for the first time.

> "It was like a blind person seeing for the first time."

In our community, we were able to visit at night, care for the sick, read, play games, listen to the radio, see each other's faces and smiles, and stay up late at night. We were happy people.

If I had to choose between electricity for one year or a new car, I would choose the electricity.

Just for one day, electricity makes a difference in my life with indoor plumbing, lights, radio, TV, and the refrigerator.

The whole house is alive.

Chapter 8

Technical Difficulties

Linemen and farmers encountered problems along the way.

Political short-circuitry

After more than a decade of battling, George Norris — along with Texas Senator Sam Rayburn, sponsored what ultimately became the REA bill in 1936.

Despite that, only a scattered few enjoyed rural electrification that first year.

Even though the government allowances were in place to begin the process of electrifying the countryside, the REA was but a temporary agency and the nation lacked the infrastructure to supply public power to all those who wanted and needed it.

Not only were operation systems scarce, so were skilled electricians. Formed out of Roosevelt's "New Deal," the REA was subject to relief agency parameters including spending one-fourth of its funds on labor.

Yet the largest obstacle was the development of an operating plan. The REA could loan money to one of three entities: power companies, municipalities, or cooperatives.

Power companies wanted some of the government pie, and exercised sizable muscle to try and get it, but the REA held firm. Likewise, cooperation with municipalities broke down.

After six months in existence, the REA still had no operation plan.

When the Roosevelt Administration was unable to reach any agreements with existing power companies for cooperation, the nation's farmers finally had the chance to show what they knew about an operation plan.

When it came to farmers, they knew the meaning and the spirit of cooperation.

One shouldn't underestimate that element of the overall scheme. Like those old barn raisings where George Norris's father met his mother, farm neighbors had enjoyed cooperative ventures for centuries, building neighborhood churches, houses, cooperative creameries, grain elevators, threshing and quilting bees, ice cream socials and covered-dish dinners.

Although its promoters were disappointed with the REA's accomplishments in its first year, rural electric pioneers did manage to hook up a few farms, and the groundwork for cooperatives was set.

Meanwhile, the rural countryside was finding a few problems with the technical aspects of George Norris's vision.

What Missouri Senator Harry Truman said about rural electrification: "I am very much interested in seeing that the farmers have as many of the good things in life as other people have."

In George Norris's Words: "If those who live in this great world of ours today could go through the same experiences and have the same fine training as the fine people of the early settlements, crime largely would be unknown, criminal courts largely would be unnecessary, and we should have in truth and in fact 'brotherhood of man and the fatherhood of God.'"

In the words of the people, here are the stories about "Technical Difficulties"....

Circle of Brightness

A series of miracles

by Herman W. Busch
Walhalla, South Carolina

Dear Lord, our Maker and Giver of all good things — of which just one of them must be the Blue Ridge Electric Co-op.

How well do I recall the day when the lights came on at our house. The year was 1941 and we lived 10 miles above Walhalla, South Carolina in an area named Mountain Rest.

My brother-in-law had wired our house, which was built in 1936 with not even a dream that we would some day have electricity.

The day finally arrived when a man came and said he was turning on our electricity. In the center of each room hung a pretty golden colored light socket and light bulb extended down about two feet, held by two green and yellow wires.

Boy, was I happy as I ran to my room to turn on my light. I had screwed a double socket into the single socket with a bulb in one and the other one left open. I reached up with my left hand and held the bulb and socket while pulling the little light switch chain with my right. Somehow I had managed to put my finger in the empty socket! Well there may not be an answer as to which comes first, the chicken or the egg, but there's no question in my mind as to which lit up first in this case. Boy those lights were bright!

> **"Somehow I had managed to put my finger in the empty socket!"**

I was in high school and I had to do all my studying at night hugged up to an old kerosene lamp, as though I was in love with the darn thing. The electric light was surely the first miracle that came with the rural electrification. We were very

Technical Difficulties

poor and all of my daylight hours after school and Saturdays were spent digging up grass or following an old mule pulling a plow in the field, or chopping wood or pulling one end of an old cross-cut saw.

The next miracle, which came several months later, after we had scraped up enough money, was to buy a used Crosley® refrigerator. Yep, and a six-footer at that — six cubic feet! That thing would make ice tea, ice cream and keep our milk so cold it would make our teeth hurt when we drank it. Six feet may sound little now, but back then, everybody cooked three meals a day and had lots of children and there was no food left over after a meal to put in the refrigerator.

A few days later after this miracle purchase was made, my mother came in one evening from milking the cow to hear a moaning and groaning coming from the dining room.

As she peeped into that room, she saw my kid brother standing there with the refrigerator door open and his tongue welded to the freezer unit with blood coming from his mouth. She ran to the kitchen where she had a kettle of hot water and dipped the dish rag in and then held it to the inside of the freezer, until she thawed him loose. That never happened again.

> "... she saw my kid brother standing there with the refrigerator door open and his tongue welded to the freezer unit."

The next miracle that came could have very well been the electric iron. This took the place of an old smoothing iron that had to be constantly returned to the hot wood stove or to the burning fire in the fireplace. This surely must have been a taste of hell to those poor women who normally had six to 12 children.

The used electric radio fell in place somewhere along

Circle of Brightness

the way. Before this, we had a battery-operated one which took only a car battery. A full charge would last for two weeks, so we bought an extra battery and it was my job to exchange it for the one in the A-Model pickup. It was turned on only at news time as the war was coming on, and maybe the "Grand Old Opry" on Saturday night.

Another little miracle came in about a year or so — in the form of a little electric fan, which my mother enjoyed so much, if and when she could steal a minute or two from her tortures in the heat.

It's for sure this next miracle measured high on the list. It was a little electric motor that sat on top of the churn with a rod extending down with a little propeller at the bottom. Boy could this little trick make butter and fast. I can remember many times pulling that old broom handle with a cross nailed to the bottom end up and down for more than an hour or more before the butter came.

My sister bought an electric stove several years after the REA arrived but our mother never really fell in love with it. She still used her wood stove until she died in 1963.

Another miracle of all miracles came in the early 1950s when a pump was installed in the well and a sink in the kitchen. Shortly thereafter came a wringer-type washer. By this time all the children were married and out of the house. Before this came along, the laundry was done outside in an old iron wash pot where the clothes were boiled in hot water with Octagon® soap. The very dirtiest clothes were washed last with Red Devil® lye added to the water. From there each piece was carried separately to the battling bench, which was a big poplar slab nailed to a tree at one end and a post on the other. Here they had the dirt — along with the devil — beat out of them with the battling stick. This was a paddle made from a board and was also kept handy with threats to "paddle" the little behinds of those who didn't mind their mothers.

Well, dear Lord, I do believe the miracle you sent us by the Blue Ridge Electric Co-op came last. That was an indoor toilet and bath. Before that, the bath was done off in a back room with a pan of water and rag. The toilet was a ways from the house in a privy. Since I am about to retire for the night by way of a nice and cozy bathroom only eight feet from my bed. I can only think how rough it would be today for this old man to stumble down the trail this cold winter night to that little shack hanging on the side of a terrace. There, I'd have to sit on that old frosty board with my teeth chattering, my bones cracking and my body shaking as that cold north wind whistled through the cracks and up that high-cut mini skirt of that old privy.

> **"I can only think of how rough it would be today for this old man to stumble down the trail this cold, dark winter night. . . ."**

I thank you again dear Lord for making all of these miracles appear to me by way of the Blue Ridge Electric Co-op.

So good night and God Bless to all as I retire for the night — by way of my nice and cozy bathroom!

Circle of Brightness

Quarantined from inspection

by Judith Church Lydick
Plainfield, Indiana

My parents, John and Sallie Church, sister Phyllis, and I moved in October of 1945 from the west side of Indianapolis to a farm north of Stilesville, Indiana ... from city life to farm living ... from electricity to darkness.

Our house had been wired for electricity but had not been inspected, thus Rural Electric Membership Corporation (REMC) would not approve turning on the power until after the inspection. In order to keep her household running, Mother borrowed kerosene lamps and flat irons from my aunts and uncles. We used a wood burning cook stove, with a water reservoir for hot water.

> **"At the end of November, I came down with scarlet fever and our home was quarantined."**

At the end of November, I came down with scarlet fever and our home was quarantined. My father was the only person allowed to come and go, and go he went because he still worked in Indianapolis.

One day, early in December, a truck pulled in the driveway. Mother looked out, recognized the utility truck as one from REMC, and began jumping up and down. Joining mother's excitement, my four-year-old sister asked, "Who is it mother? Santa Claus?"

I remember mother saying, "It's just as good!"

Her celebrating was quickly subdued when the inspector said he could not come inside to inspect the wiring because we were quarantined — he had small children at home too. The look on my Mother's face must have shown true disappointment — the strain of being shut in the house nursing a sick child, caring for an active pre-schooler, continuing to run the home without electricity . . . . city life was looking more appealing again.

The inspector looked at her and said, "Well, I have never found anything wrong with this electrician's wiring, so I'll inspect the barn and the outside wiring, If it looks O.K. I'll pass the house.

He did. We passed!

The next night, on a cold dark night in December, every light in the house and the barn was on when my dad came home from work.

It was a good Christmas that year . . . I was over my scarlet fever and we had lights on our Christmas tree!

Circle of Brightness

The house with no lights

by Cora Stephens
Rudyard, Michigan

I am writing this in memory of my late husband who passed away in 1994 after a long illness at the age of 79. We were married in 1940.

When I was growing up, I often wondered who lived in the house without lights. I used to see a lamp in the house.

After we met, he told me how as a young boy he would read by that lamp. The reason their lights were never on was they came from a small country town and during the Depression, his family lost all their money.

It was a happy time when they got their lights turned back on. I still have that little old lamp he read by. He read a lot even until he passed away.

War delayed mom's 'fridge'

by Robert Rice
Monon, Indiana

I was in sixth grade when the lights came on at my parents' farm home in the late summer of 1941.

We waited with great anticipation for the electric contractor to get to our house. He was working from one house to the next, coming down the road.

My mother was excited to have electric lights in the house, which meant she would soon have running water in the kitchen. It meant for dad, he could do away with the gasoline engine that pumped water for the livestock.

One of the first things I remember Dad converting the old battery radio to electric. This did away with the wind charger out in the yard, which had kept the batteries charged for the radio.

Then the war started.

Because the war came along, mother did not get an electric refrigerator until they became available after the war. Dad was first to get a new refrigerator at the local new appliance dealer when one came in, and a new electric washing machine, and best of all, we had a new bathroom.

Circle of Brightness

Who needs electric anyway?

by Henry L. Hill
Glenview, Illinois

Who needs electricity? That's what my mother said. This story begins when my parents, who lived in Chicago, discovered Beaver Island in 1917. They spent practically every summer there, as have I for many years.

Geographically, Beaver Island is part of an archipelago of islands at the northern end of Lake Michigan. Approximately 16 miles long and six miles wide, it is the largest island in the Lake. Early settlers were mostly of Irish descent, making their living fishing and lumbering. Probably its greatest contribution to history came in the late 1840s when it was occupied briefly by a splinter group of Mormons led by James Jesse Strang.

Our cabin had no electricity, we were rural and remote. Yet, somehow, lacking electricity didn't seem important. We were no different than anybody else in the country. Everybody pumped water and if our well ran dry we had plenty of water in the lake. Nobody else had inside plumbing.

One of those beautiful pull-down lamps with an alabaster white shade circumscribed by a brass ring dangling a fringe of red and clear glass prisms — a prize of antique collectors — lighted the sitting area of our small cottage. A glass pedestal lamp graced the kitchen table by whose light we could also play cards after dinner. And in the corner, over the stove, was a bracket lamp to see what was in the pot.

Later one of those gas lanterns, with its soft, hissing sound, lit up almost the entire cottage. It was also a very comforting light on the way to the outhouse. No washing machine or dryer for dirty clothes? Just get out the tub and the corrugated, zinc washboard and then hang the clothes on the

Technical Difficulties

line. Toast for breakfast? Sorry, no pop up toaster, take off one of the lids on the range and put that little square rack over the fire, just don't let the toast burn. Too bad, no electric juicer, apply some muscle to that glass reamer instead.

Of course, at the end of the summer we always retreated to Chicago, where everyone had electricity, so we never confronted a frozen pump or had to slog through the snow and sit in a freezing outhouse to relieve one's self. Who needed electricity, indeed!

That was mother's thinking even after electricity finally came to the entire island in the 1960s as Top O'Mich, our rural cooperative, extended service from town to the end of the island.

> **"I'm happy with my kerosene lamps and I don't need electricity, thank you."**

Signing up members, Top O'Mich even offered to install service from the main line to the house without a charge. However, when the company's representative knocked to offer mother that great opportunity, she simply said, "I'm happy with my kerosene lamps and I don't need electricity, thank you."

And that was mother's position until she died. After her death, we finally obtained electricity. Running the line was no longer free, but now we have the benefits of electricity — and inside plumbing. Wow! Thanks Top O'Mich.

Circle of Brightness

Timing never right

by Arnez Gans
Octavia, Nebraska

Bumpety-bump-bump!

Halfway down our wooden stairs in our farm home, I came down on both my knees. As a result there were two big holes in my good, long cotton stockings. I was a very dedicated high school senior at the time, and the embarrassment I felt among my family members was worse that the pain. This accident could have been more catastrophic, but fortunately, the kerosene lamp that I was carrying was still in my hand.

As I came to an abrupt landing at the foot of the stairs, my way to see in the dark was still glowing. It was my nightly ritual to study upstairs at my own little writing desk with the lamp directly in front of me. (My optometrist tells me too much eye strain in dim lights caused my myopia).

"Mom! When are we going to get electricity? Then I wouldn't have to carry this smelly old lamp wherever I go," I blurted. But I really knew why. Our country, at that time, was involved in World War II, and many things were scarce and rationed. I was very much aware that more modern conveniences were difficult to come by.

Prior to this time, President Franklin Roosevelt was spending time at his little cottage in Warm Springs, Georgia. Thanks to his ingenuity, he came up with the idea of "Bringing light and power to every nook and cranny in America."

Of course, we at our farm home in Nebraska were not aware of this wonderful thought of lighting the rural areas as well as the cities. As a result of this idea, on May 11, 1935, the Executive Order was signed by President Roosevelt that brought the Rural Electrification Administration into being. Thanks, also, to our Nebraska State Senator, George W. Norris, who played an important part in rural electrification for out state. This truly changed the lives of all rural people.

However, on our little farm home in Butler County, it was a very long time coming! When the war finally ended, so were my high school days, as the dates of the beginning and end coincided with my school years. Many of our neighbors on the surrounding farms already had electricity. Since we lived quite a long way from the main graveled highway, with a long lane, it was difficult to obtain the necessary poles and other equipment. This is what I remember my parents were told. As a result, I believe that we were the very last in our immediate community to see the light.

> "When the war finally ended, so were my high school days, as the dates of the beginning and end coincided with my school years."

Good news finally came, and I remember our farm home being wired for electricity. My mother was thrilled as she chose pretty light fixtures for the various rooms of the house. The old battery radio was discarded for an electric one. Our old ice box was replaced by a new electric refrigerator. An electric stove was installed, and the old cob and wood cook stove was put out in the wash house. Oh, yes, it was in this wash house where the old gasoline-powered family wash machine was replaced by one with an electric motor. That was indeed a major improvement for I remember so well the difficulty mother had in getting the old thing to start with a foot pedal on washing day and there was always a lot of laundry with a family of six. These, I believe were among the first conveniences that we enjoyed along with the new experiences of turning the lights on.

'Robbing Peter to pay Paul...'

by James F. Jackson
Carlisle, Indiana

John and Mary had a large family and several years ago had bought a washing machine run by a gasoline engine. The washer was old and hard to start. One spring morning John was preparing to plant corn and Mary called to him.

"John, you gotta fix this washer. It won't start."

John checked the engine. It seemed a new spark plug might fix it. He didn't want to go to town for a new plug. Maybe a plug from the Ford V-8 would work. The plug from the Ford did fit and the engine ran.

"It's fixed now. I gotta go plan corn."

That afternoon, Mary delivered some eggs to town, but did not replace the spark plug in the Ford.

That night she said, "John, you gotta fix that Ford. I drove it to town and it made an awful noise all the way there and back."

John looked under the hood of the Ford and said, "We gotta get an electric motor for that washing machine."

The day lights didn't come on

by Allen Weaver
Culver, Indiana

I remember when the Rural Electric Membership Corporation (REMC) lights came on or should I amend it by saying — "WHEN THE LIGHTS DID NOT COME ON!"

It was in the late 1930s, and still during the Great Depression. My father borrowed the money to wire the house, and to buy lamps and appliances. We were required to have the wiring checked before the power was turned on.

The REMC worker came out to check the wiring using a generator. The generator was only meant to check the wiring with no lights or appliances on. The worker pulled the starting cord on the generator. It started for a short while, and coughed and sputtered and quit.

My mother threw up her hands and screamed.

> "It (the generator) started for a short while, coughed, sputtered and quit."

"Now we will have to borrow more money to re-wire the house."

It turned out O.K. because her little boy (me) ran from room to room and turned on all the lights and appliances he could find.

Circle of Brightness

Of lights and war

by Hilda Burke
Myrtle Beach, South Carolina

I was born in the early 1930s, so I remember gas light very well. The mantle that was put over the gas jet was very delicate. So you had to be careful as they were quite expensive to replace. Two pennies at that time was a lot of money.

I use to like to watch the gas lights flicker as I fell asleep. My mother left the light on low for us at bed time. Then turned it off when she went to bed herself. This was before World War II came to England.

Years later when our home was blasted from the bombing all around us, our house wasn't safe to live in. So we moved to a house farther away from the docks, which the Germans were bombing.

> **"Years later when our home was blasted from the bombing all around us, our house wasn't safe to live in."**

The house we moved into already had electricity, so we thought we were rich, even though we couldn't keep the telephone that was already in the house.

Even though the house had electricity, the street lamps were still gas. We knew it was time to go home, when the gas lighter with the long stick came to light the lamps. We played in the street close to the house enjoying the last of the daylight, before the lamp lighter made his rounds.

It was nice later on, to be able to use an electric iron instead of the flat black iron which we had to heat on the coal fire.

Technical Difficulties

My mother cooked on the coal fire before we had electricity. We still laugh about the time she was cooking sausages for my dad's dinner, when the soot fell into the frying pan. So she took the pan and flipped the sausages over the back yard wall of the chimney sweep, who lived a few doors away. We often wonder what the chimney sweep thought, when he saw the sausages flying though the air. He was supposed to have just cleaned our chimney, hence the flying sausages.

During war's "blackout" we had no lights in the street. If we went out after dark, we'd have to take a flashlight and keep in pointed at the ground so "Jerry the German" wouldn't see the light if they flew over.

As Vera Lynn sang in those days, "When the lights Go on again, all over the world," we too couldn't wait.
I think that's why I love the lights of Christmas, and the beautiful home displays, with so many lights. It's a far cry from the blackout.

Circle of Brightness

Explosion ruined feast

by Paul R. Honan, M.D.
Lebanon, Indiana

Boone County Indiana was the first county in the nation to benefit from rural electrification in 1935. My parents' farm home, five miles from the county seat of Lebanon, was one of the first new electric lines. It was an exciting event for my parents, sister and me.

A pre-owned Delco® 32-volt direct current electrical system had replaced the kerosene lamps a couple of years earlier. A one-cylinder gasoline engine with a flywheel powered the Delco® generator.

The bank of nearly worn-out, glass-enclosed storage batteries could hold only a partial electric charge. They provided not-so-bright lighting when the generator was turned off.

> **"There was a flash of light as all four light bulbs in the ceiling light fixture exploded."**

A real electrician from the city rewired the old farmhouse for the new electric power. The brightness of the new 110-volt city lighting in the country was dazzling and wonderful the day the lights were turned on.

The following Sunday, my mother prepared a country dinner for guests after church. The dining room table was filled with platters of fried chicken, mashed potatoes, homemade bread and jams and other homegrown specialties.

As our family and guests were ready to be seated at the dining room table, my mother decided we needed more light in the room. She flipped the light switch on the wall. Instantly, there was a flash of light as all four light bulbs in the ceiling light fixture exploded. Tiny pieces of glass showered down on every platter of food and glass of water on the table.

Then we realized that the lights had been turned on in all the rooms previously except the dining room. The unintentional 220-volt power supply into the dining room was too much for the 110-volt light bulbs.

After throwing out all the food on the table, my mother dutifully prepared another meal for the guests.

After the initial disaster, we settled down to enjoy the new rural electric power. A shiny, chrome-plated General Electric® toaster and a tabletop Westinghouse® radio in a wooden cabinet were luxury items.

Circle of Brightness

City bride learns rural ways

by Alice Laski
Phillips, Wisconsin

Here I was in 1948, a young bride of three years, on a farm in Wisconsin my husband and I bought. I was accustomed to electricity having spent several years working in Milwaukee.

Here, I had a wood cook stove, a gas iron, a gas-powered wash machine (which tested all my patience), and hot water from a copper boiler heated on the cook stove. Without any running water, it had to be carried in. There was a gas engine on the pump, which rebelled in our sub-zero winters.

Then came rumors and news that we would be getting electricity in the countryside. I started to dream. Next we saw the power lines going up. Wasn't too long later, we had an electrician wire our buildings. Our first large appliance purchase was a refrigerator. It was to be used for our first child's appearance that year. No sour milk for our baby!

The day finally came when the lights came on. Everyone as far as the eye could see had their houses and barns lit up. What a welcome sight. We toured the neighborhood that night, just looking at the lights.

There was no end to what electric power did for farmers, to make life easier, and happier. Every building was wired now, meaning many new conveniences. With a milking machine, we could milk more cows, and add income. With running water in barn, it saved many hours pumping water for cows. We now could put a bathroom in the house. Talk of feeling good. We no longer had to heat water on the stove, or carry water in. The kerosene and gas lanterns and lamps were discarded.

God Bless our hero George W. Norris, who had the insight to get electricity to the countryside by his legislation, and the coops to make it affordable for us. Thank you.

Fire took electrified home

by Phyllis Biggs
Mecosta, Michigan

We used to have to do our homework by gas lanterns. My brothers and I were in school. My mother used a cook stove (wood) to boil water for baths and laundry. You had to start supper right after breakfast to make sure it was done. We took baths in a tub on Saturday night to be ready for church on Sundays. We also had an outdoor toilet and my dad put lime in it quite often to keep the smell down.

In the 1940s the electricity came. Mom threw out her cook stove and scrub board and dad took the gas lanterns out to the barn. Mom got an electric stove and a wringer washer. It sure saved on the knuckles. We could also take a bath with clean water for all. We all celebrated when the lights came on. It was great. It was sure rough without it but never having had electricity, we didn't realize how rough it was.

Then five years later we had a fire. We were all away when it happened. There were two families in our town with the same name as us. The firemen went to the first one in town and by the time they got to our house in the country it was too late. We lost everything even our house.

Dad had to go to Flint to get a job so we could survive. I moved closer to town with a house that had no electric. It was like starting over again.

When you tell your children and grandchildren about it, they think you came from another planet. I remember it well as I lived it.

It was no picnic.

I'm sure glad we have it now, as you don't realize how important it is till you don't have it.

Wired house waited 30 years

by Robert E. Smith
Broken Bow, Nebraska

The house that I have called home for more than 45 years was purchased by my parents Vince and Mabel Smith back in 1934. That might seem like strange timing but the folks had been school teachers and dad — in 1934 — was drawing about $150 a month as the superintendent of Elwood, Nebraska schools. Not much since he was raising a family and subsidizing a small farming operation, but affluent compared to most who were depending entirely on agriculture.

The quarter section with the farmstead was considered a showplace with relatively new buildings, built in 1918. I have often wondered at the faith and hope of those people who included wiring for electricity in the plans for that large two-story house. Could they have imagined that it would be nearly 30 years before any juice would go through those wires? But the great Depression, terrible drought and World War II all came into play.

> "I have often wondered at the faith and hope of those people who included wiring for electricity..."

The early builders lost the place and did not live past the mid 1930s. This opened the doors for the Smiths to move in. Dad added some more land to the farm but by the end of the war, we still operated with one 25-horse power tractor and draft horses.

It was now 1946 and new things were in the works, including a 110-volt home light plant. There was a waiting list as demand was high for those up-to-date innovations.

It seemed that new and better tractors were likely to take over the role that horses had long played. A buyer arrived and much to the dismay of this writer, Dad sold a large share of our work string. The deal required they be delivered to the stock yards at Berwyn, 13 dirt road miles away.

The sad day arrived, and dad and this very small 15-year-old saddled up to drive the ten-head herd of my four-footed friends to the new owners. Some 25 miles and many hours later, after darkness had fallen, we reached the top of the section hill a mile from home.

A marvelous sight met our eyes. The home place to the west appeared to have a halo over it. The crew had been there to install the new plant. Looking back, it is hard to imagine how a few 40-watt bulbs could make such a difference, but it was plain that those were not kerosene lamps. We covered that last mile in record time, to enjoy mom's waiting supper under the bright lights, and the relief of not cranking the cream separator. The new life had arrived.

The move toward REA was already in the works in our area. Apparently it was 1944 when dad came home from the first of hundreds of meetings with the news that Custer Public Power District was born. He said some day those little holes in our walls with the wire ends would be more than a nuisance for the paper hanger. By the time the lines arrived in our neighborhood in 1951, the great blizzards of 1948-1949 were history. I graduated from high school, the Korean war arrived, and I was spending two years with Uncle Sam. These were all big events in my life, along with the day the lights came on.

Now some 50 years and three generations later, the Smith family expresses our appreciation for the vision of George Norris, the many years my dad spent as a board member of the Custer Public Power District, and to all those who had and continue to have a part in brightening up rural America.

Chapter 9

Hallelujah!

To many rural residents electricity was a heavenly miracle.

Hallelujah!

Then . . . slowly . . . the lights really did come on.

Once the funding mechanism and the proper operational plan was in place, the REA grew to maturity and nobody took more pride in this government program than the nation's rural countryside. A farmer wrote Norris, "I congratulate you on this movement . . . it is the only way that the younger generation will be interested in establishing houses in the rural communities."

Shortly after the Norris-Rayburn Rural Electrification Act took effect, the REA became overwhelmed with applications for rural service. Farmers took the initiative and formed cooperatives with their neighbors. When a potential line was formed, the process began — usually for a $5 deposit. Most often, a rural resident applied for a loan which included information about the operation and mapping details of the cooperative area, as well as price schedules.

Among the first items purchased for the rural farmstead were lights, an iron and a radio. Running water and indoor bathrooms came later.

Still the power companies intervened. Some became proactive in efforts to destroy cooperatives before they had the chance to begin. They waged misinformation campaigns, and infiltrated cooperatives by suddenly building a single "spite" line to an advantageous member of a cooperative in an attempt to break it up by skimming the choice members.

In its first year, the REA resolved much of its operational procedures, and by 1938, those policies and procedures were ironed out enough to begin seeing the results as they began to light up the farms.

REA officials worked hard to process loans faster and faster while engineers developed better line-erecting techniques and more cost-effective methods of delivering electricity to the rural countryside.

By July of 1939, more than 100,000 miles of new lines reached more than one million rural residents and by the end of the year, the REA estimated that rural households with electricity had risen to 25 percent.

On the first day of 1940 — after four and a half years in existence — the REA had financed more than 180,000 miles of new lines with another 80,000 either under construction or planned.

During that boom period, many rural Americans experienced what may have been one of their most memorable days ever.

What President Jimmy Carter Said About The Day The Lights Came On: "... I think the best day in my life, the one that I remember most vividly — with the possible exception of my wedding day — was the day they turned the lights on in our house."

In George Norris's Words: " Men have been establishing homes, building factories, harnessing streams, building canyons, conquering the earth and the skies. And yet rich are the people in this world to whom trees, and water, and growing things bring hope and happiness.' "

In the words of the people, here are the stories about the day they received lights, "Hallelujah!" . . . .

A day to remember!

by Ruth M. Martin
Reynold, Indiana

I was living on the farm in White County, Indiana in 1942 and what a wonderful day when the lights came on. We looked forward to this day for months.

My father and a neighbor wired the house with a ceiling light in every room and one or two outlets in each room of our eight room house. They cut the hole in the ceilings for the light fixture and we kids would look through the hole and called to each other on the lower floor. We were cautioned not to step on the hole and fall.

> **"We kids ran around the house flipping on and off the light switches to see if the lights worked in each room."**

I was attending a one-room school about a mile from our home. The school had no electricity before that day either. We had the outdoor toilets, one for the girls and one for the boys. We also had a big round coal-burning heating stove, that had to be fired by the teacher. One of the older kids had the job of carrying in coal and cobs from the outdoor coal house and helping the teacher carry out the ashes. These coal ashes (or cinders) were put on the paths to the outhouses, so they wouldn't be muddy in bad weather.

Oh what fun the day we came home from school and found the power had come on at our house while we were gone.!

We kids ran around the house flipping on and off the light switches to see if the lights worked in each room. The one

Hallelujah!

we liked the best was in the upstairs hallway at the top of the stairs. We could turn it on or off either upstairs or downstairs. How exciting! And how did that work?

Of course we didn't have any electric appliances yet on that day. We still cooked on the old wood and cob cooking stove and heated the water there. Our house had a small hand pump that pumped rain water from an outside cistern well. We kids had the job of pumping water to a tank which was fastened at the ceiling. Water ran by gravity through a faucet in the bathroom, which had a tin sink and a tub on four legs. No stool, of course.

At that time we had a refrigerator that ran on kerosene. It was more modern than the icebox which our neighbors had. We could make homemade ice cream in this refrigerator by using a mix with our own whole milk, stirred a couple of times while it was freezing, and presto, we had ice cream.

We also had no toaster; if we wanted toast, we held it over the hot stove. Usually, we had bacon, eggs, biscuits and jelly for breakfast. No coffee maker either, we made coffee with a percolator placed on the hot stove and let it "perc" or boil for a while.

We had no electric washer, but we did have a Maytag washer which was run by a gasoline motor. This washer with a wringer was placed in the wash house. We used two galvanized tubs for the cold rinse water. The hot water was carried to the wash house from the house.

What a wonderful day when the lights came on!

Circle of Brightness

A new kind of warmth

by Madelyn Hatchell
Conway, South Carolina

Before electricity came to our house, I remember having to warm a blanket by the wood fire and then running to the bed with it so it would keep us warm until the bed was warm.

There were three of us girls and we shared the same room with two beds. We would all three sleep together in cold to keep warm. I was around 10 when we first got electricity in about 1946.

Electricity changed a lot of things in our life. We did not have to go to town to get ice to keep things cold. We bought a refrigerator. If you never had a refrigerator before, it was better than Santa at Christmas. You could look down the road to see if your neighbors were home, because you could see the bright lights shining.

Those were years that I will always remember and cherish, for as long as I live. You had to cook with a wood stove and warm by the fireplace.

Hallelujah!

With the turn of a switch. . .

by Leora Williams
Richmond Center, Wisconsin

What a great day when the lights came on. On cold winter days, I disliked pumping water for the livestock. chickens and hogs. You had to pump and carry water to the house for cooking, drinking, cleaning, laundry, bath and etc. You had to pump water in the cooling tank to cool the milk and you milked cows by hand. You had to drive horses on the hay rope to unload hay.

Now all you have to do is turn on a switch.

With the turn of a switch, the elevator takes hay to the mow.

With the turn a switch, motors pump the water, heat the water for livestock, start the milkers, transfer the milk to the bulk tank and cools it.

With the turn of a switch, you have very good lighting, an electric washer, and dryer. The is no more carrying milk and butter to the root cellar to keep cool. The refrigerator does the work, and with the freezer, you can have your fresh meat, etc. You have running water, and hot water for laundry and baths. With the turn of a switch you have electric heat fans, vacuum cleaners, etc. Electricity helps mow lawns.

Name it, electric does the job. I can name many, many more. I appreciate electricity, a great invention.

It was a great day when you flipped a switch and the lights came on.

Circle of Brightness

'Isn't this swell!'

by Janeen Jackson
Roundup, Montana

Bessie Von Olnhausen's favorite expression was: "Isn't this swell!" That is just what she thought when electricity came to their farm. Charles and Bessie Van Olnhausen farmed beneath the Crazy Mountains on the southern end of the Golden Valley County, southwest of Ryegate, Montana.

The first cook stove Bessie used was a wood burning stove, then in July of 1928, they bought a three-burner Coleman®, that burned white gas. Excitement really filled the air when on May 26, 1937, they ordered a wind charger that would bring them electricity; but they found themselves still waiting for it to arrive by December 16. Bessie noticed the wind and wished they had the wind charger. Impatience was noted in her diary entry.

> "Bessie noticed the wind and wished they had the wind charger. Impatience was noted in her diary entry."

The day before Christmas they got the wind charger and had electric lights at last. It only had enough power to run lights and when there was no wind they were really dim. They also ordered a new radio the following Saturday, an important item for them. It was run on a dry and wet cell battery though and had to be taken to Ryegate every three months for recharging. It was easily overcharged also. Things were looking up now. They had lights and a radio to catch the news and weather.

The wind charger, charged a six-volt battery that stood

Hallelujah!

about three feet tall, about a foot square and held water. Some of their neighbors had 32-volt chargers, but they didn't get them until after Von Olnhausens.

After coal oil lamps, the lights were much brighter and their son Elmo was glad to be relieved of his job of polishing chimneys It was no easy task as they were easily broken.

The six-volt wind charger was a disappointment, and they wrote the company about it. No matter, it brought a welcome change. In May of 1945, Charles replaced it with a Koehler® generator that produced 110 volts, and that was a real upgrade. Then toasters and irons could be used.

On January 20, 1950, the electrician came from Melville to wire their farm house and connect them to Montana Power. Their electric bulbs were flush against the ceiling instead of hanging from a wire in the middle of the room as some were. Some fixtures had a little knob on the side for switching on lights, but theirs had a pull chain. Bessie had fun shopping for those modern light fixtures.

They had jumped the gun, thinking electricity was a long ways off, and had prematurely upgraded to propane appliances by then, but when the electrician came he also brought Bessie a new washing machine.

The gasoline washer with the engine that was hard to start, was retired. No more metal flexible exhaust hose poked through a hole in the wall to vent leaking fumes outside; fumes that seeped in anyway.

The new washer was the real high spot when the lights come on and it was heaven.

Circle of Brightness

A new start

by Amanda Rose Frierson
Olanta, South Carolina

It was a new beginning when I got lights because it was tough before.

Now I enjoy life.

I can see and listen to TV musicals on my TV. I can listen to the radio, talk on the telephone and enjoy electric fans and the freezer, as well as running water.

If it wasn't for electricity, I wouldn't have the things I have. I thank the Good Lord for blessing me in 1956 for allowing us to get electricity.

It was like a new beginning.

Daily life changed for better

by Goldie Kluck
Schuyler, Nebraska

I am a great-grandmother now, but when the lights came on years ago, I was a young wife and mother filling kerosene lamps and washing glass globes. As a young teenager at my parents' farm home, it was my daily chore to keep lamps in readiness. I have one faithful kerosene lamp I kept as a reminder, days before rural electrification.

Electric lamps are wonderful. So are electric stoves — no fuel to carry in, baking done tasty good with a timer. Years ago, you couldn't always depend on the old cook stove.

The electric cream separator was a great asset for my husband and I, as we milked many cows by hand. We separated cream from milk bucket-by-buckets, turning separators by hand. That cream check meant a lot to people. With it, they could buy many grocery items, gas for the family car and a few clothes.

The electric iron took the place of my three sad irons which I heated on the cook stove covered with a pan.

Many other electric appliances and farm equipment made work easier and profitable.

It was a great pleasure to set down in evenings with a big bowl of popcorn and enjoy the radio or television.

The alarm clock rings, and tells you to get out of bed. It's going to be another busy day. Much of it made possible by electricity.

Thanks to the Rural Electric Administration and to George W. Norris, the U.S. Senator from Nebraska — the legislative father of rural electrification.

Life got better

by Jannie Mae Floyd
Lake City, South Carolina

I was a farm girl living in Lake City, South Carolina. Before electricity, we used kerosene lamps, lanterns and Aladdin® lamps to see and study by. My brothers, sisters and myself would play yard games, play with our pet billy goat or listen to the phonograph for entertainment.

Our lights were turned on in the latter part of 1949. The day the lights came on marked a transitional period in my home and community.

Electricity gradually changed everything tremendously. We could see after dark. The light from the light bulbs was so much better than the light from lamps and lanterns.

> **"Electricity gradually changed everything tremendously."**

My father bought an electric refrigerator and we did not have to buy the 100-pound block of ice each week to keep our food from spoiling. This eliminated the weekly purchase from the ice truck. After packing 100 pounds of ice in a wood box, we had to wrap the ice in croaker sacks and put our food around it.

After the electric refrigerator, next, we got an electric stove, so we no longer had to cut wood to cook our food. Before electricity, we used wood to heat our home and to cook our food. It required lots of wood to do this, so we had to cut, haul and stack it daily.

Then came our electric, ringer-type, washing machine. This was a lifesaver. We did not have to tote water to wash, scrub or rinse our clothes. Although, we still had to hang the clothes out on lines to dry we no longer had to scrub clothes clean on a rub board, or boil them in a wash pot outside.

We could also iron our clothes with an electric iron without getting soot on them. Before, we had to heated our iron in the fireplace or on the stove or heater. The iron had to be cleaned real good before ironing to keep from getting soot on the clothes. As you can imagine, that was virtually impossible.

Before electricity, our bathroom was an outside toilet and a Number 3 wash tub. We drew water from a well with a bucket tied to a rope and toted it to our house. The well was also used to store our milk and keep it cool.

My father used to have to walk for miles to listen too "The Grand Old Opera" on the radio.

With electricity, now we could all listen to the radio at our own home.

All these changes occurred due to the discovery of electricity. It all seemed so amazing and still does.

Electricity is taken for granted today. If people had to grow up the way that I did, they would understand how electricity has simplified our lives and made the world more enjoyable.

Thank God for giving man the inspiration to discover such wonders as electricity.

Circle of Brightness

Wife gave electric ultimatum

by Josephine Nowak
Medford, Wisconsin

    I was born in 1917, growing up in a one-parent house. Dad died when I was only seven. We had no electricity on the farm, so we carried water in pails, and filled the cattle tank by by melting snow.

    In 1938 I went to town to work.

    I was married in 1940 and in 1943 my husband decided he wanted to buy a farm, but it didn't have electricity.

    I agreed only if he would hook up electricity.

    For two weeks we lived there before it was hooked up, and it was awful. I washed clothes on the washboard for two weeks, for my husband and I and our two children, plus diapers. We milked 20 cows by hand.

    I was never so happy in my life, when the lights went on. I remember the day well — October 27, 1943 from Clark REA.

A life improved

by Irene Robison Rust
Fort Wayne, Indiana

We looked forward to having electricity but we didn't realize then how it would change our lives.

We had everything we really needed. A base burner, coal-heating stove, a cook stove in the kitchen and a square Maytag® washing machine that ran on gasoline.

With no refrigeration, we knew if we had mashed potatoes for dinner, we'd have potato cakes for supper!

We looked forward to the day the wiring crew would get around to our house. It took some time.

Well, it happened. One afternoon when I arrived home from school. I was greeted by one of the workers who said, "Come here, I have something to show you." I followed him to the living room where he turned the switch on a lamp mounted on the wall.

Yes, the lights came on!

From then on life changed. First there was an electric range, then a refrigerator, and lights in every room — and a bathroom with warm running water! These are just a few ways our lives were changed for the better.

God Bless the day rural America got electricity.

Circle of Brightness

We've come a long way, baby

by Berniece Zimmerman
Broken Bow, Nebraska

We've come a long way, baby!

Yes, growing up in the 1930s, I can relate to the good old days, somewhat, but for the most part, I'd rather enjoy the comfort and a conveniences that rural electrification has made possible.

We lived in the Nebraska Sandhills in a four-room house in the wide open spaces. The winter of 1935 was one of the coldest on record. The heating stove burned cow chips by day but was banked and re-fed with chunks of coal during the night. The bed was moved into the living room dangerously close to the stove. Morning found the frost from our breath on the top six inches of the comforters.

My father got up first to start the kitchen range. The dipper would be frozen solid in the water pail and the bread frozen hard as a brick. We had a deep freeze in those days and didn't even know it.

> **"We had a deep freeze in those days and didn't even know it."**

We had a kerosene lamp, mounted on the wall with an attractive round reflector to direct the light where it was needed most.

Mother had a square tin contraption with wires on one side for making toast on the range. With no way to control the heat, the bread usually burned or dried out.

One place where we lived, the water ran from the windmill into a shaded tank in a building and then out into the stock tank. The cream can milk and butter were kept cold in the water there. Another place had a wooden barrel to serve the same purpose for smaller containers, like the yummy chocolate milk which was my special treat.

A gas refrigerator was purchased after I left home but I never got to enjoy this convenience until we got REA in Custer County.

In the summer of 1951, we bought a refrigerator. The store agreed to keep it until we got the promised electricity. We had our house wired anticipating the day we could put away our kerosene and gas lamps. That day came on December 19, 1951. I came home from teaching school, flipped the kitchen switch and wow. We finally had electricity.

The previously purchased refrigerator was delivered. I received an electric iron and waffle iron for Christmas. An electric toaster was soon purchased which made a nice golden treat.

Over the next few years, an electric well replaced the windmill and a milking machine took the place of milking by hand. An electric welder made it possible to repair most equipment breakage at home.

Yes, rural electrification added a big plus to more convenient and comfortable country living. We don't realize how dependant we are on electricity until the power goes off. We can't get a drink, cook, iron, sew, flush the stool, operate the radio or T.V., use a number of shop tools, operate electric irrigation wells, and many other modern conveniences.

Yes, we've come a long way, baby!

Circle of Brightness

Electricity extended day

by Mrs. Lindsay L. Wood
McCormick, South Carolina

In the 1930s before electricity came to our home, we used kerosene lamps, a wood stove for cooking and heating, buried large blocks of ice in a box filled with sawdust to prevent fast melting, brought water from a spring, and kept butter and milk in gallon jars lowered into a neighbor's well.

When Aiken Electric Co-Op started running electric lines in rural Edgefield County, my Daddy, George A. Gilchrist, started work with them in 1940. He wired many homes and churches all over the county. I was 12 and wondered what possible changed electricity could make in our lives. Did it ever!

We started night services at church and even visited neighbors at night — something we never did before. We stayed up later at night, being able to see so much better to read and sew.

Since we lived one half mile off main Highway 23 between Edgefield and Modoc, we were among the last in our community to have electricity turned on. In September 1941, the lights came on!

Daddy loved music of all kinds and the first thing he bought was a beautiful floor model Philco® radio. My mother raised the roof because she said there were many other conveniences she could use. That was the first and last argument they ever had, because the next item bought was a Westinghouse® refrigerator. It made its own ice!

Then in December 1941, my Christmas tree had the only string of lights in the neighborhood. Some neighbors warned, "Be careful, they may set your tree on fire." The Philco® is still in the family and the tree lights are safely stored.

Thank God for electricity

by Mrs. Kelly McCormick
Loris, South Carolina

It was December 16, 1942, the day our lights came on in our home. We have always lived on a farm, got our water from a well for all of our use including for all the different animals we had on the farm. We studied by a oil lamp all of our school days.

The lights have changed our life more than any other change that has come to our homes.

When our lights go off, we all panic until they are restored. Everyone is worried and they try to fine someone with a generator so all the food in the freezer doesn't spoil.

We all owe the repairmen and the forefathers of rural electrification a lot.

Thank God for the knowledge he gave man who invented electric that has given us so much happiness.

Electricity brought hope

by Jean White
Collbran, Colorado

In 1937 during the Depression years, we were fortunate because we lived on a farm and could produce virtually all of our food. But times were hard and we did without many of the better things in life.

One day a man came to the door to be welcomed warmly by my parents. After talking business for a while, the man opened his briefcase and produced papers which they all signed. Of course, I —being only 10 — did not understand it all but I knew it was important because mother and dad spent hours pouring over the Montgomery Ward® Catalog.

> "I knew it was important because Mother and Dad spent hours pouring over the Montgomery Ward® Catalog."

A few days later, as I returned home from school, I found a man at our house working in the walls. Mother explained to me that he was "wiring" our house for electricity.

Men came and set poles and ran wire from them to our house. Then came the day the electricity was turned on. Miracles of Miracles. We had electric lights!

As I went from room to room pulling the string to watch the lights come on (we did not have wall switches) Mother told me not to turn the lights on unless I needed them since it would run our bill up.

Now I could do my homework in my room instead of sitting at the kitchen table near the kerosene lamp.

We were all excited. I did not see how things could be

Hallelujah!

better. We had it all! Now there would be no more filling kerosene lamps, trimming wicks, or cleaning lamp chimneys.

Several days later, the order from Montgomery Ward® arrived. It contained a large radio. Now we could listen to our favorite radio programs without fear of running the battery down.

Next, came the wringer washing machine. It was placed in a corner of the kitchen and needed only to be rolled to the center of the floor to be plugged into the light socket. Now mother would no longer need the washboard.

Mother's eyes sparkled when she saw the electric iron. I know how she hated keeping a fire going to heat the sad irons even in the hot summer months. She now had only to set the ironing board under the light socket and plug the iron in.

Best of all was the refrigerator. Now we would not need to carry our milk, cream, butter and eggs to the cellar to keep them cool.

There was much discussion about indoor plumbing since we now had power to run a pump to bring the water into the house. That, however would have to wait until more money was available. But really, who cared? The path to the outhouse was merely an inconvenience. We had much to look forward to. Our entire world, present and future changed the day the lights came on.

Chapter 10

Relief From Drudgery

Electricity revolutionized farm kitchens across the nation.

Improving quality of life

George Norris knew first hand the drudgery associated with rural farm life and not just in his home state of Nebraska. Through his travels in establishing the Tennessee Valley Authority, he saw what wash day meant in a sweltering Alabama afternoon and he knew how difficult it was to milk a dozen cows twice a day during the blizzard season in North Dakota.

Those in the Roosevelt Administration continued to promote rural electrification efforts because many believed that affordable electricity would improve the standard of living and the economic competitiveness of the family farm.

Clyde T. Ellis, the first general manager of the NRECA, wanted to be at his home when the REA came to his home in 1940.

"When they finally came on, the lights just barely glowed. I remember my mother smiling. When they came on full, tears started to run down her cheeks."

The REA gained momentum with men like Norris, and Ellis, who knew the monumental change lights were bringing to the rural countryside. From the lineman in the field to the highest levels, the REA realized the significance of their efforts as evidenced by their slogan, "If you put a light on every farm, you put a light in every heart."

Quite simply, the REA changed lives.

Some historians believe the REA may have been the most important legislation the government ever enacted on behalf of rural Americans.

Some politicians concede that while the REA started slowly, it did do much to help a nation recover from the ravages of the Great Depression.

Politics and history aside, anyone who ever lived on the

farm without electricity would tell you the rural electrification was one of the most memorable things that ever happened in their lives.

All over the United States, the people staged ceremonies and mock funerals to paying their last respects to the kerosene lantern and celebrating their new electrified lives.

What a Tennessee farmer said while giving witness in a rural church: "Brothers and sisters, I want to tell you this. The greatest thing on earth is to have the love of God in your heart, and the next greatest thing is to have electricity in your house."

In George Norris's Words: "From boyhood, I had seen first-hand the grim drudgery and grind which has been the common lot of eight generations of American farm women, seeking happiness and contentment on the soil."

In the words of the people, here are their stories about their "Relief from Drudgery"....

Circle of Brightness

Electricity brought respite

by L.R. Rebenitsch
Fort Rice, North Dakota

We had the house wired for electricity long before we received it in our area in 1951. Then came electricity! What a great day when the lights came on!

With lights in the house there was no more kerosene or gas lights. And no more pumping water by hand. And the bathrooms were such a blessing! Garden hoses now took care of the water carrying.

Of course all the electric things were acquired as we could afford them. Electric irons replaced the old sad irons or gas irons.

Iceboxes gave way to refrigerators and later freezers. Milk machines allowed the ordinary farmer to increase his herd.

Electric stoves cooked and baked your bread. What a relief in the summer's heat not to have to fire up the range.

> **"What a relief in the summer's heat not to have to fire up the range!"**

Before electricity, life on the farm was hard but families were solid. Everybody had to work — we didn't have time for much mischief.

To bring back memories, kerosene lamps lit up our homes but not too well. Every night the lamp chimneys had to be cleaned, the wicks trimmed, the bowl filled — a smelly job. If you were lucky you had a gas lamp (mantle) which hung from the ceiling. They made a much brighter light than the kerosene lamp but the bugs and moths attracted by the light, soon broke the mantles. Some had lamp shades to push the light downward but moths still found their way to the mantles.

Relief from Drudgery

It seemed the water pail was always empty and the slop pails always full. You needed to dump the slop pail outside but the dishwater went to the pigs so we had two slop pails. Some people had windmills to pump water but most had hand pumps that took energy — to fill your pail full was a cold job. Sometimes the pump froze up so it meant bringing out the ever-present tea kettle of hot water and thawing it out.

The cook stove had a reservoir that needed constant filling to supply warm water at all times. The tea kettle also needed filling.

When the wind didn't blow, we had to pump water for the horses, cows, pigs and chickens. They seemed to drink more when you had to pump by hand. Then there was wash day which meant carrying tubs of water plus a boiler for the stove. It took all day to scrub and rinse and wring the clothes. The hand wringer was screwed onto the tub and usually the kids helped turn it. The wash water was then carried out and used in the garden. We had no lawns and few flower gardens back then.

The cook stove needed a constant supply of wood or coal. Then the ashes had to be cleaned out — a dirty job. The furnaces and heaters also needed wood and coal. Then there were the outhouses — usually a bit of a walk away from the house. At night you used a flashlight if you could afford one or a kerosene lantern that gave a very dim light. A night chamber or the potty was used for the elderly and the children. That had to be carried out every morning to the outhouse.

Another chore was separating the milk after hand milking. It took a lot of energy to get the separator up to full speed and then the children could keep it going. Of course a lot of water was carried to clean up afterwards.

Looking back 50 years, my husband and I still marvel at the changes electricity brought to us.

Circle of Brightness

Then the work load lessened

by Mrs. John Mayer
Medford, Wisconsin

A child of the 1920s, I remember well no electricity. We lived one mile outside of Medford, Wisconsin, city limits but were able to attend city schools. We walked two and a half miles each way. I loved school for the lights, warmth and indoor plumbing. Never missed a day. Lights at home were a kerosene lamp for the house and a lantern for the barn.

Our outhouse was between the house and barn. No running water. A gas-powered engine and a pump-jack pumped water for the farm animals. Water for the house was hand pumped, carried by the pailful to fill the tea kettles and kitchen wood stove reservoir. This was for drinking, cooking, cleaning, bathing in a washtub and washing clothes. Our heat came from a wood cook stove.

I remember tea kettles, flat irons for ironing and a copper boiler filled with water heating for washing clothes.

I remember my mother used a scrub board for a family of 10 children. Later she had a hand-powered wash machine with a hand wringer. Us 10 children provided the washing machine's hand power.

> **"Us 10 children provided the washing machine's hand power."**

Clothes were hung outdoors in summer, and on wooden clothes racks by the stove to dry in winter.

I remember making toast by laying a slice of homemade bread on the stove. We also used a long-handled holder for holding slices of bread over hot, live coals. We also heated a curling iron in the globe of the kerosene lamp and singed lots of hair.

Relief from Drudgery

In 1947 I married. My husband bought a farm in Chelsea, Wisconsin, Taylor County, two miles west of Highway 13. Lake Superior had electricity along Highway 13 and one mile in, to the east and west. R.E.A. came later for the farmers.

Our farm had lights installed in 1941 before we bought it. It consisted of four bulbs in the barn, one in the hayloft and a 100-watt bulb yard light. The house had four bulbs, one in each room. We had cold running water. We still had the outhouse.

We installed electric water heaters, added a bathroom, and an electric wash machine. We still dried clothes outside in the summer and hung them on wood racks in winter. The electric dryers came later. We burned fuel oil for heat.

Today we seem paralyzed when we lose electricity.

Circle of Brightness

Electricity expanded life

by Mildred L. Kearney
Kearney, Nebraska

With the flip of a button the lights could go on in every room. What a delightful treat!

The gas light fit its purpose, but only lit one room and was also difficult to carry from room to room. We lived at the south edge of Custer County, Nebraska, but our power came from Dawson County because there weren't any close neighbors in our county.

The big yard light lit a big space. How nice to see after dark to get in and out of our car. When leaving for the night, to go somewhere it was usually left on and was always comforting and helpful to find when returning home.

Our boys had a basketball hoop on the garage, so the yard light made it possible to throw some baskets even after dark. The yard light also gave us light to see to go to the barn to milk cows. Now we could even switch on lights inside the barn, instead of milking by dim lantern light.

> **"... the yard light made it possible to throw some baskets even after dark."**

Electric light also made it easier to play games, read, or do sewing and mending because you didn't have to be so close to the table holding the old gas lamp. It also made for a lot less strain on the eyes.

Washing wasn't that much different after getting electricity at first, but oh — the joy and help it was when we got our first automatic washer and dryer! The dryer made winter washing much easier, but on nice days we still took

Relief from Drudgery

advantage of our Nebraska winds to dry our clothes on the line. The electric iron replaced the flat irons and gas irons. It was so simple just to plug it in and have varying degrees of heat. That was back in the days before wash and wear, when almost everything needed to be ironed before being worn.

The hot water heater made washing dishes, clothes, and our baths so much more enjoyable, and required less work and time. A wash tub isn't quite as accommodating to bathe in as a bath tub when you're an adult.

Also convenient for us was our deep freeze where we stored beef, pork, chickens, and even vegetables and fruit. Having the freezer also created more time because we didn't have to can so many foods to keep them good. We lived eight miles from town so it was nice to have food on hand.

We don't realize how many joys and conveniences we take for granted until one of our Nebraska storms takes away our electric power. Thank you Dawson County Public Power District.

Circle of Brightness

Affordable luxury - finally

by Theresa Bogner Montee-Nelson
Dickinson, North Dakota

Dad was five when he came here from the Banat region of Hungary in 1903, a German Swabian. The Bogner homestead is located at Lefor, North Dakota.

Dad was very interested in new things. Our first electricity was a cord, a bulb and a black six-volt-battery, then came the 32-volt and many batteries — a wind-powered charger

At last, after the second world war, we got rural electricity! We lived 25 miles from Dickinson, N.D.

There was no more hand-cranking the cream separator. There were no interruptions in a radio program. We had a refrigerator. After the war, these things were not too easy to come by. We used the Montgomery catalog a lot. We had electric lights in the yard, no more kerosene lanterns.

Such labor-saving devices, made us even more proud to be an American.

It was an affordable luxury.

Thank you, all of you, who made this possible for all of us.

No more community well

by Judy N. Broughton
Cross, South Carolina

What was it like before electricity came to our home in Cross, South Carolina? As for our family, which lives in the rural area, there was no running water, there were seven families who lived on the hill, who had to use one hand pump with a pitcher. The well was drilled some time ago, but the water was cool and clear.

During those days kerosene lamps, lanterns, candles and light woods, were used to make a blaze of flight in the chimney, so my sisters and brothers could study our lessons for school when my parents didn't have money to buy fuel.

The lights did wonders in the area, where there were no lights. My father bought an iron, radio, washing machine, and brought running water in the house after our house was wired.

Our church was wired also, because my father usually lit the mantel lamps when there was a service during the night, because it takes a little know-how, in lighting those lamps.

The electric cooperative brought lights in our area in the early 1950s.

Since the light came through, many other electrical things were purchased. I could go on and on to tell how wonderful the Co-op light was to all in the community. Every year the company give out some useful things.

May God continue to bless rural cooperatives and its members.

Daily routines changed

by Eleanor G. Thompson
Onida, South Dakota

I certainly remember the day the lights came on in our home in Summit township, Sully County, central South Dakota. It was on a bright May day in 1950. It was perhaps comparable to the automobile when it became available to pioneers of the oxen teams or the horse and buggy mode of travel.

Many of our usual ways of doing our daily routine were changed at once. But some had to wait until finances would permit.

> "Many of our usual ways of doing our daily routine were changed at once."

Of course the first big improvement was when evening came and we did not have to sit by the kerosene lamp — which was on the table — to read, write or study. As soon as possible the barn was wired and those with livestock found it much easier to care for their animals than it was by lantern light. Milking machines and electric separators came later.

The housewife found laundry day was much easier when the washing machine exhaust did not have to be put out the window nor did she have to wear herself out kicking the starter.

There were no wrinkle free garments then. The ironing was usually done with sad irons which were heated on a stove and necessitated replacing the other one as it cooled.

The electric cook stove with a timer and temperature control and not having to replenish the fuel source at intervals was absolutely marvelous — especially when baking angel food cakes from scratch — which I did for customers. We also had a large flock of laying hens and kept a light with a timer in the hen house. Our battery radio was replaced with an electric one and much later a television was purchased. The electric refrigerator was certainly welcome as it saved food and we could always have a cool drink.

Prior to the R.E.A. when we would return from our weekly trip to Onida on a Saturday night I would hurry to the well for a fresh bucket of water. It was four years later when we put down a cistern and got a water pump, so we could have a bathroom. No more traipsing to the old outhouse in snow or sleet, and to the discarded Sears® or Montgomery® catalogs.

Always an avid reader from childhood, I was pleased to be able to read in bed occasionally — especially when not feeling well. In later years when I drove back and forth to work in Pierre. I had an electric heater on my car.

Thank God and the REA for making a better and more comfortable life for all of us. The service is still very good. The Oahe Electric Coop at Blunt has dedicated employees.

Circle of Brightness

Electricity changed our pace

by Marjorie Crick
Soldiers Grove, Wisconsin

Life in the country without electricity was much quieter and slower paced than today.

When dad wanted to get electricity on our farm he was told he would have to pay for several miles of line himself. But he decided it would be worth the expense. He had to pay a certain amount each month which was more than the electricity that we used until the line was paid for.

True life was much different when we received electricity. I am thankful for the many advantages electricity gives us. Yet some of the old ways were enjoyable.

> **"True life was much different when we received electricity."**

Let me give you a bird's-eye view of what changed with electricity.

I'll start with Saturday because it was a day to prepare for the week ahead. Food was prepared for the Lord's Day. No fast foods then you know. And we pumped and heated water for the Saturday night bath in the old round tub. No daily baths in those days. Then we had to lay out clothes for church and go off to bed — so blow out the kerosene lamp.

Sunday was a day of worship, relaxing and playing games. Only the necessary work was done.

Monday meant bringing in more water and heating it on the cook stove for it was wash day. Clothes were washed in that same old round tub with a wash board and hung out to dry, even if it meant you had to freeze dry it. How funny it was to see Dad's long underwear frozen stiff like a board.

Relief from Drudgery

Tuesday was the day to do the ironing. One needed to heat the flat iron on the cook stove and iron clothes while another iron was heating then switch to the hot iron and continue until done.

Since we had no refrigeration we cooled our food in the well house in cold water. We thought it was great to get hard butter when mom got a refrigerator.

When mom baked, it meant beating a cake or batch of cookies by hand with a wooden spoon. Then you needed to fire up the cook stove and wait until the oven was just the right temperature for your baking. We always had fresh baked bread because mom baked it herself. It took a lot of bread to feed a family of eight and sometimes a hired man,

Dad and my older sister milked eight to 10 cows by hand. One time when my sister was sick, I decided I would help in her place. I milked one cow while dad milked all the rest.

Circle of Brightness

A celebrated improvement

by Dorothy Waldron
Chadron, Nebraska

In June of 1955, we put away the kerosene lamps (and are now quite amazed at the prices such things command in antique shops). Before the new REA line arrived from two miles away, the old farmhouse had first to be wired for electricity.

As farm income allowed, 110-volt appliances began replacing the 12-volt equipment, which had been powered by a wind charger set up in the yard, the tall metal tower with long wooden blades generating electricity which was stored in large glass batteries in a small building. Of course, when the wind failed to blow there was soon no electricity available for anything except a minimum of lighting and in June we were without even that as lightning struck the tower earlier in the summer and had blown up a number of the batteries.

With reliable power from the REA, the old battery-operated radio was used no more and we no longer had to restrict listening time because of the drain on the storage batteries.

This was especially nice when doing the weekly two bushels of ironing using the new electric iron instead of sweltering in a kitchen heating flat irons on the wood-fired range. Once, I brought an outdoor thermometer to check the indoor temperature and found I was working in 125-degree

> **"Once, I brought an outdoor thermometer to check the indoor temperature and found I was working in 125-degree heat."**

heat! An electric motor replaced the noisy gasoline engine on the washing machine and housework was much easier with a vacuum cleaner.

A new and larger refrigerator replaced the small 12-volt appliance and we could also discontinue use of the huge kerosene-fueled refrigerator kept on an enclosed porch for storage of milk and cream.

The men's work was easier too. Yard lights and lights in the barn did away with the uncertain light and constant fire hazard of kerosene lanterns. An electric welder and various tools such as electric drills and saws made routine repairs and general upkeep simpler and we found that the welder could even be used to thaw frozen water lines, so no more carrying buckets of boiling water to frozen hydrants at the stock tanks in the winter.

Our children were just reaching school age in 1955 and now had good lighting by which to do homework or play games in the evenings. They also felt more secure since each one now had a light available in their bedrooms if they awoke at night.

Yes, "When the lights came on" was a time of general celebration for our family. We still had our work to do but now it went more easily, we were less tired at the end of each day and as parents we had more time to socialize, to read to the children or play games with them, to entertain friends or tackle home improvement projects.

Children no longer hesitated to invite their friends to visit, they did lessons more easily with adequate lighting and even began an after-school routine of listening to favorite children's programs on the new radio.

Chapter 11

Enchanted Water

Running water and electrification made the ice man lonely.

Circle of Brightness

The quest for water

Running water, both as a consumable product and as a source for energy, occupied much of George Norris's career in politics. He became chairman of the Committee on Agriculutre Forestry in 1920 and a year later, water became a critical concern when the government asked for bids to lease Muscle Shoals, a series of whirlpool rapids on the Tennessee River which were a natural source of water power. When automaker Henry Ford submitted his bid, it raised a question concerning the benefits of private interests, as opposed to government ownership, and stirred debate over the merits of the federal government's role in such development.

Because he opposed the highly popular Ford, some communities burned Norris in effigy, but he continued his push to use natural resource for the public good and Ford eventually withdrew his offer.

In 1926, Norris introduced legislation providing for the operation of a dam at Muscle Shoals, but no action was taken.

Three years after that, the senator authored a resolution to provide for the national defense by creating a corporation for the operation of the government properties at or near Muscle Shoals. This cleared both branches of Congress but was vetoed by President Hoover.

During the 12-year TVA battle, flood waters flowed down the tributaries of the Tennessee River and ravaged property and lives along this and other waterways.

On April 11, 1933, Norris finally succeeded in getting legislation favorable to the development of the Tennessee Valley.

Norris always regarded the TVA and the REA as necessary twins. After favorable TVA legislation passed, Norris set his sites on extending electricity to the farms.

Enchanted Water

What Sen. Lister Hill said about George Norris decade-long effort to establish the TVA: "It was a titanic job he undertook. To those who would continue on the trail he blazed, this is a truth to remember. George Norris never asked what was feasible. He asked what was right. Ne never asked how long and hard the road ahead would be. He saw a distant goal and knew he had to start without delay . . . George Norris fought that the farmer and the housewife might participate in the modern world."

In George Norris's Words: "Every stream in the United States which flows from the mountains through the meadows to the sea has the possibility of producing electricity for cheap power and cheap lighting, to be carried into the homes and businesses and industry of the American people."

In the words of the people, here are their stories about "Enchanted Water" . . . .

Like a fairy tale, water came

by Clarice Sabata
Valparaiso, Nebraska

It was hard to know how electricity would really affect our lives. But the things I heard people say this wonderful invention could do was like a fairy tale dream to a 10-year-old girl like me.

Our whole family was excited when the crew came by and put in poles and strung the wires. We could hardly wait for the power to be turned on. Could there really be such a thing as lights at the flip of a switch and instant running water from a faucet?

I usually helped dad outside while my younger sister helped mom inside. I never liked housework — especially doing dishes. But I promised mom when we got electricity I would do dishes every day without being asked. What a joy it would be to have not only water — but hot water from the faucet.

> "What a joy it would be to have not only water — but water hot from the faucet."

The day the crew came by and actually turned on the power, dad was in the field and came home after dark. We thought we'd really surprise him. We lit the kerosene lamp and waited. When he came in the house we flipped the switch and yelled "SURPRISE." But he already knew by looking at the transformer so the surprise was on us.

But what a wonderful difference it made in all our lives. No more studying by lamp light. We could see so well to read while mom sewed. And even more important to kids — we could stay up later since we had light.

Another advantage (or maybe it was a disadvantage to some) was being able to take a bath anytime we wanted or needed. No more Saturday night bath with one tea kettle of hot water per person added to the same tub everyone else used. Poor dad — he was always last. Now he could have clean, hot water too.

Even more important on a farm was the outside hydrant. Cattle tanks could be filled every day with good, cold water, no longer were we dependent on the wind. And a long hose would water the garden so its produce would sustain us all winter. Eventually an irrigation well would guarantee water to corn crops even in the driest summer.

I'm glad I remember life before electricity. It makes me appreciate having electricity that much more and gives me wonderful memories to share with my grandchildren.

Recalls water system

by Chet Bednarski
Mayville, Michigan

June 18, 1938. I remember that date well. Although I was only nine, I watched the light bulb slowly come to life in our lives.

My three brothers Louis, Steve and Walter attended the open house in Ubly. On display were all types of appliances for the home and time-saving devices for the farm.

The first light lines put up had only two wires, which made it very vulnerable to the elements (storms, high winds, ice, etc.) Gradually, service became much better with more substations and improved lines.

My father had our old house wired for electricity for some time before the great event. The electricians that did the work had a hard time leading the wire from room to room because of the construction of our home.

The first conveniences invested in were a Zenith® radio and a Surge® milking machine. We milked about 10 to 20 cows twice a day, so electricity was a real labor saver. Gradually, more appliances were added when money was available.

Funny, how we take things for granted. Like flipping a switch for lights or turning a tap for hot or cold water, setting a thermostat for heating or cooling. I remember trying to read at night with a kerosene lamp, heating water on a stove for a bath or trying to keep warm standing in front of a potbelly stove.

Yet, what I remember best was the water system — running water in the home, how wonderful that was.

The best glass of water ever

by Carol (Hollender) Furman
Mantello, Wisconsin

Because of electricity I have a memory to pass from generation to generation.

As a child we lived on a farm. My brother and I both had certain chores to do. In those years it was not uncommon for your grandparents to share part of the house, so wood and water was a chore used and needed for both families.

I was eight years old when electricity came to us and we were all excited. To be able to flip a switch and a light in the ceiling come on was truly a revelation. We maybe had two plug-ins in each room and only downstairs. We didn't get plugs in the bedrooms upstairs until years later.

As in these days, the kitchen was the hub of everything that happened. To turn the faucet on and get hot and cold water enough to fill a wash tub and take a bath was great! Of course the hot water was due to that upright thing in the basement that was on a board, because we had a dirt floor in the basement. My father took my brother and me downstairs and explained to us about the pump and the water heater. Even after explaining all this I wasn't really sure how it all worked, but it did.

The point of my story is when Christmas came that year, which was always the best time, usually what my Dad got me was always special. I was a typical daddy's girl. He left the living room for a few minutes and when he came back he had a tray with glasses for everyone — filled with water. Call it happiness or joy to the maximum but that glass of water was the best gift we could have ever gotten. I know it was the best gift I ever received then and up to the present. We all knew how much having electricity meant to each of us and it was my dad's greatest gift to us all.

Our first tray of ice

by Shirley Lail
West Columbia, South Carolina

In the early 1950s most everyone in our neighborhood in Hickory, North Carolina already had electricity in their homes except for just a few families. Our family was one of those. We were very poor and were renting a house for $8 a month. The house did not have any closets, no bathroom and no running water.

We walked about a half-mile to one of our neighbors to watch the "Amos and Andy" show on TV and all the other shows that came on Thursday night. This was usually the only night of the week we would go there because we felt like intruders — even though they were kind enough to invite us.

This same neighbor (whom we thought was very rich) also had a refrigerator in their home. She was nice enough to sell us a tray of ice for a nickel a tray. The ice made our tea so wonderful to drink.

> "She was nice enough to sell us a tray of ice for a nickel a tray."

We could only get ice from the ice man about once every week or two and when the big block of ice melted, we were without ice for the rest of the time. A nickel sure seemed like a lot of money to us back then since it took a whole day of picking cotton to earn about 38 cents.

But then the big day came when our landlord had our house wired for electricity. We didn't have any electric lamps but oh how beautiful that light bulb in the center of the ceiling looked to us. No more studying with kerosene lamps.

And then the most wonderful sight I ever saw was our

second-hand refrigerator that came on and made us not one, but two glorious trays of ice.

Today, I have two refrigerators — one of them has a water dispenser and a choice of crushed ice or cubed ice — and I thoroughly enjoy these features, but it still doesn't make me as happy as when I saw the first tray of ice in our home in the 1950s.

One good thing about having ice in the home, I don't have to walk a half mile to get a tray of ice and then walk home to find out my nickel's worth of ice is two and a half cents worth of water by the time I carry it home in the hot sunshine.

Circle of Brightness

Running water changed lives

by Darlene Doyle
Newport, Nebraska

"Run, get a pail of water."

The four children in my family were separated by 10 years, two older and two younger, one of us was usually handy to fetch water. The water supply came, teeth hurting cold, from the windmill and pumped into a small tank in the milk house.

Many's the time I'd lean over and sip a cool drink. Cream and milk sat on raised bricks, to just below the surface. The overflow traveled out to the garden tank where the gold fish lived. The dipping bucket was not easy for a child to lift, and I'm sure our pails were never brim full.

Once inside, the bucket was placed on the wash stand by the door or dumped into the reservoir of the kitchen range. This last action meant one probably needed to, "run get another pail."

> "But it was the running water that changed our lives and the face of the farm."

Electricity came to our Box Butte County farm in 1950. Electric lights were a novelty and a blessing after the lamps and lanterns. No more wick trimming and chimney cleaning, or pumping up the lantern. No more scary, dark trips to shut the chicken house door, or milking by lantern light.

But it was the running water that changed our lives and the face of the farm.

Thanks to the REA, we received hot and cold running water!

We had water to lovingly wash the best china after company dinners.

Enchanted Water

We had water to rinse the first radishes and onions from the garden — water to nourish that garden.

Before REA, water was siphoned from the garden tank and ditched down the rows. After installation of a hydrant, Mom would sit in the evening, hose in hand, by the garden or flower beds, spraying the precious water on her thirsty plants. Cool, green grass graced the tall, white house and invited one to sit in elm tree shade.

The process of carrying water from the milk house was reversed, and now we took a bucket of hot water out to Dad. This he used to wash the electric cream separator.

Somewhere we needed to find space for a bathroom. Half the large kitchen was sacrificed and remodeled for the glistening white fixtures. No more tin-tub Saturday night baths by the old range, or summer showers under the big barrel full of sun-heated water. No more fast trips to the outhouse at the end of the yard. An addition to the house made more kitchen space and room for the washing machine and tubs, a lavatory and shower. Mom delighted in spraying cold water through the window onto Dad as he showered off the field dirt.

From a time when every drop of water was used and reused (such as Dad's bath being used kids bath water) to a time when it was OK to enjoy a cool shower after a hot, itchy day shocking oats or to splash an unsuspecting family member with cold water from the hose.

All thanks to the Rural Electric Administration.

Circle of Brightness

The chamber pot incident

by Betty Dembny
Gaylord, Michigan

It was about 1940. I lived on a farm in Posen, Michigan with my parents and 11 brothers and sisters. I remember how horrible it was to be without hot water. No inside bathroom, but an outhouse. But somehow we all survived without electricity. One cold day in January we were lucky enough to have the electric poles and wires come across the rolling fields and roads to our house. We were poor — I don't know how my parents could afford to pay for it.

We kids walked two and-a-half miles to school every day and then back. It didn't matter — rain or shine, or the coldest winter day. On one very cold day, we arrived home from school and to our surprise we had electricity.

We were overjoyed. We started playing with the "On" and "Off" switches, in every room and outdoor lights, having a great time with it, our mom didn't even holler at us.

We had a night (chamber) pot upstairs in our bedroom and every morning, it was someone's job or duty to bring it down and empty it. For some reason, this very day it was forgotten, until we returned home from school so my brother, Ali, decided he would do that job.

As he was walking outdoors carrying the pot, he yells to us "Put the light on, put the lights on" and was not paying attention where he was walking, he stepped on a patch of ice, he slipped and fell with a bang, spilling the contents on himself.

It was so funny to us kids, we never let him forget it, we teased him "put the light on — Bang."

That was our first day of having electricity.

Oh what joy!

The faucet made us kings

by Carolyn S. Smith
Monticello, Indiana

I was about seven, living in a rural area of Indiana. Mother lit oil lamps every night. Sometimes the lamps smoked and mother turned down the lamp and wiped the chimney clean of the smoke film. Just before electricity came, Father bought an electric radio and table lamps. One looked just like the old oil lamp. While the family enjoyed the newness of all the bright lights and the barn dance on the new radio, Father told Mother to turn the lamp down because it was smoking. Mother reached for the new lamp which looked like the oil lamp, out of habit, before she realized what she was doing. Father laughed and said. "Isn't electricity great? No more smoking lamps."

It was great to only have to flip a switch to light the stairway upstairs. Even greater was running water in the house.

No more going out in cold weather and pumping water. We carried many buckets of water to the big washtub which Mother heated on the gas stove. She heated water this way for our baths and laundry on Monday. I woke up every Monday to the sound of the gas motor on the old wringer machine, going putt, putt, putt. Sometimes it would backfire. BANG! If it was a nice day, Mother hung the clothes out on the clothesline. If not, clothes could be found hanging in almost every room of the house. The new electric washer and dryer made Mother's wash day faster and easier. We heated water in tea kettles for dishes, and rinsing. To just turn on a faucet for hot or cold water, we felt we were living like kings, a life of luxury.

I wouldn't like to go back to those days, but I'm glad I was able to experience the "good ol' days. It makes me appreciate more of what I have today.

The miracle transformed us

by Donna Franklin
Rapid River, Michigan

Electricity changed life in our rural area in many ways. When electricity was installed at our place, it was a miracle to us kids, and nearly so to adults, I suspect.

Before we had electricity, every drop of water used in the house had to be hand-pumped and carried in from the well. Watering livestock in winter was a back-breaking chore, since the water had to be hand-pumped and conveyed to the barn by truck, or by hand when the snow got too deep. Vegetable gardens had to be located close to sources of water and lack of irrigation dictated the size and types of crops we could grow.

The coming of electricity to rural areas made it possible for farmers to raise larger and different types of crops, enlarge livestock holdings, and expand all facets of farming which increased the farm family income.

> **"My first memory of the miracle of electricity was turning on the kitchen water tap and watching the water flow."**

My first memory of the miracle of electricity was turning on the kitchen water tap and watching the water flow. I couldn't believe it. I thought it was some error which would quit any moment, so I kept going back and turning the water on and off until grandmother made me quit.

The electric water pump brought us an abundance of water for home and outside use. An outside tap was installed for the needs of the livestock, the garden and the farm machinery. Hoses and ditching conveyed the water to the places it was needed.

Electricity changed our recreation. We could read after dark. Each family member soon acquired our very own radio for our room. Before that, radio usage was limited because of the expense of replacing the batteries.

The pinnacle of luxury was reached at our home with the installation of a three-piece bathroom complete with running hot and cold water. No more dashes to the outhouse on frigid winter mornings. No more washing hands in icy water fresh from the well before breakfast. No more heating water to wash up before walking to the end of the lane to catch the school bus.

I couldn't conceive of any glamorous movie stars living any better than we did once the indoor bathroom was installed.

Chapter 12

Social Illumination

The radio was just one of the changes brought by REA.

Circle of Brightness

Empowering rural America

Not only were the lights turned on, but electricity enlightened a rural countryside in ways some people never imagined. They went from a time of inconvenience to a time of comfort. They went from hand and horse power to electrical motors.

Electricity extended day, secured the nights and filled lives with a new sense of joy. In a speech to Nebraska farmers, George Norris said REA electricity would be like having new hired hands around the farm — one to help the farmer and a hired girl to help his wife.

"This hired girl will work 24 hours in the day and the longer and harder she works the more she will enjoy it and the less will be her rate of pay," Norris said.

"This hired man will saw the wood, light the barn, milk the cows, fill the silo, water the lawn and the stock. He will even irrigate the farm and do all these and innumerable other things without complaint and with complete satisfaction."

Leisure time increased. It encouraged reading, stimulated renewed interest in books and in stories, in music and the arts. It meant deliverance from the dark.

The TVA reported electrical lights added two to four waking hours per day. Another study showed electricity added 91 eight-hour days per year. Yet another study determined that with electric water pumps and the availability of running water, rural farm families saved more than 30 minutes per day. Rural residents saved 20 days per year using electric washing machines instead of the previous method of tubs and scrub boards.

Electricity promoted sanitation with electrical washing machines and refrigeration, thus discouraging disease, improving health conditions and lessening the fatigue of farm women. Better lighting prolonged eyesight, allowed safer working conditions and provided immeasurable psychological benefits.

The radio brought entertainment, delivered news and other important information and brought a rural countryside up to speed, furthered education and exposed people to worlds previous unattainable.

What President Jimmy Carter said about his rural upbringing: "My life on the farm during the Great Depression more nearly resembled farm life of fully 2,000 years ago than farm life today."

In George Norris's Words: "I could close my eyes and recall the innumerable scenes of the harvest and the unending, punishing tasks performed by hundreds of thousands of women, uncomplainingly and even gayly and happily, growing old prematurely; dying before their time; conscious of the great gap between their lives and the lives of those whom the accident of birth or choice placed in the towns and cities."

In the words of the people, these are their stories about "Social Illumination" . . . .

A journey from 'second class'

by Elaine Frasier
Max, Nebraska

Come with me, if you will, to another time in our rural Nebraska's past. When day disappeared into night and the light with it. When an evening in a nearby town, illuminated by lights powered by generators, was a visit to a wondrous place. When kerosene lamps in rural homes made dim attempts to keep the darkness at bay. When a trip outside to shut the chickens for the night was a journey in the dark. And, a trip to the outhouse was a real challenge if the flashlight batteries failed. Here was this vast community of darkness as dusk fell. And it remained until the first streaks of morning sunrise appeared in the eastern sky.

Rural Electrification arrived in Southwest Nebraska in 1950. My most profound memory as a 10-year-old living on a farm with my family was the night we took the journey.

We were in the family car returning home from town. My parents had picked me up after school and we had traveled on to town so my mother could do her weekly trading. That involved taking cream and eggs to the produce station, and buying the week's groceries at the store.

My dad stopped at the local livestock auction to see what was happening. By the time we turned onto our lane, it was past dusk, with no moon; the stars were becoming visible in the night sky. As we neared the turn into the yard, I remember my mother taking a quick breath and saying, "Look at all those lights!" Tiny, glittering lights dotted the horizon to the north of us, ahead of us and to the south. My dad stopped the car and we simply sat there and gazed. Those magical lights were the yard lights of neighbors living near and as far away as we could see. We couldn't imagine there could be so many of

> "At that moment, my normally predictable parents decided to do something totally unpredictable."

us out there. At that moment, my normally practical, predictable parents decided to do something totally unpredictable.

My dad turned the car around and drove down country roads, from one yard light to the next, my dad or mom identifying who lived there. We continued into the next county, making a large circle, just totally amazed at the distance we could see those lights from our home on a hill.

Looking back and analyzing the situation, I think I understand the cause and effect. When the farming community was first connected to the REA lines, lights, a few appliances for those few plug-ins and perhaps a water pump were the extent of usage. A minimum rate was charged and those few lights didn't use enough kilowatts to fill the bill. Leaving the yard light on for several hours each evening was simply using what was already paid for.

But, just maybe, I like to think, those practical, predictable farmers were just proclaiming to the world, "We're here and we're not second class citizens any more!"

The great equalizer

by Rebecca D. Bolin
Rowesville, South Carolina

Edisto Electric Cooperative brought electricity to our home in Cattle Creek Community, Orangeburg County, South Carolina in 1938. The night our current was turned on, we switched on every light in our two-story home! Next we rode around our community where most of the people illuminated their entire homes too!

Electricity was the best advantage that we have ever had for our community. It turned night into day!

Before Edisto Electric Cooperative switched on the lights, all labor was time-consuming and energy-intensive. Most work was done by hand and animal labor, with minimum help from gasoline engines. We used open wood fires in fireplaces and stoves, including cook stoves for warmth, laundry, cooking, and heating water. Cutting and hauling wood for these fires were time and energy consuming occupations. Water was pumped or drawn by hand from outdoor wells, which along with ice from icehouses several miles away, sometimes provided refrigeration. House and wildfires were extinguished without running water. We ironed with cast-iron irons, heated by the fireplace. We had no electric yard lights for security purposes.

On the farm, we milked cows by hand, and used hand-operated cream separators and churns. Irrigation of crops depended on the weather, or was done by carrying water in heavy buckets. We used mules to grind sugar cane, and boiled the syrup, using lightwood fires under kettles. Wood stoves heated the chick brooder. Hog butchering took a lot longer, since we had no electricity to heat water or to power a sausage stuffer.

Social Illumination

During camp meeting at nearby Cattle Creek Campground, pinewood knots on fire stands were used for illumination. Oil lamps lighted the church, tabernacle, and individual tents. Schools, churches, and public places used wood heat, and oil and kerosene lighting.

> **"Now rural residents have the same opportunities provided by electricity as those in the city."**

That was the way it was. Now, rural residents have the same opportunities provided by electricity as those in the city. We are able to use electric ranges, refrigerators, freezers, both inside and outside lights, automatic washers, dryers, electric irons and other labor-saving appliances. We can have indoor plumbing, with hot and cold running water.

We now have electronic communication, entertainment and education. Computers and telephones connect us to the world.

Electricity enables citizens to use schools, churches, and other public buildings day and night, in all seasons of the year. Rural fire departments can fight fires, with greater success. Rural industries and farms flourish. Rural residents can keep house, farm, and have careers, all in the same lifetime.

The day the lights came on, Cattle Creek Community came out of the dark ages into the 20th Century, the Age of Technology.

Light ended fearful nights

by Mary Catherine Woodbury
Effingham, South Carolina

Thanks to God our Savior Jesus Christ, for bringing electricity to our lives.

Our home was always cold, because we only had warmth from a kerosene heater and wooden stove. Although we had a wood stove and kerosene heater, it was always bad for us because my brother Fulton had severe Asthma. He couldn't take the smell of wood burning, and the smell of kerosene. So during the winter, he suffered a lot with Asthma, and I could tell the hurt that my family carried.

We had no hot, running water. During winter, the three of us sisters and three brothers had to heat water on the wood burning stove. But we always thanked God for what we had.

Not only did my brother suffer from the cold, but my mother suffered badly from Rheumatoid Arthritis.

We had an outhouse for the bathroom, so when night came we always had to use a night pot. Which I can still smell today. I remember always catching a cold because of the cold weather. My two sisters and I always slept together, to keep each other warm. My three brothers did the same as we did.

We had no television and radio. But we did have each other. I always remember praying to God for us to have a better life.

Then our house became rat-and-snake infested. Snakes and rats came in our house at night and moved around the house during the day. Rats lived in our bedrooms, and snakes lived in the bedrooms and kitchen. I remember my mother's scream when she saw a snake in the kitchen as she was putting wood in the wood stove.

We were not able to stay up late at night because we

needed to conserve on oil and candles. Many nights my sisters and I would talk in bed without light — always afraid of rats and snakes.

We also didn't have any lights for our Christmas trees. But we did sit around the tree always wishing we had lights.

When we first got electricity, we were able to socialize with the people. We began to have dinners and started staying up at night. Plus Christmas really felt like Christmas as we were able to see our neighbor's homes all lit up.

Plus our home became safe and more secure, we were able to see a difference in my mom and my brother's illness. We were able to go over to our grandma's house and stay for longer times. Our church begins to have nightly songs.

Being black, the Ku Klux Klan stopped visiting our house so we didn't have to be afraid of them any more. Dad had a night light so we saw who they were. The same people who used to talk to us during the day light.

All together we became a hopping community. We first got electricity when I was about six. Then we moved into a new house that was built with a bathroom.

> **"Being black, the Ku Klux Klan stopped visiting our house so we didn't have to be afraid of them any more."**

We praised God, for all he had done for us, through electricity.

Circle of Brightness

Moving up in the world

by James R. Todd
Conway, South Carolina

In the summer of 1943, Horry Electric turned on electricity on RFD in Loris, South Carolina. Before we got electricity all we had for lights was a kerosene lamp and a chimney that we burned wood in. That gave us some light. We had no refrigerators and nothing that used electricity.

We had to buy blocks of ice and wrap it in a toe sack, put it in a tub, and put sawdust around it to keep it from melting. After a long time we finally got a radio so we could listen to the "Grand Ole Opry." Later on, when we could, we got an electric iron instead of one that you had to put corn cobs in to light and make it hot.

> **"And when we finally got a refrigerator, boy we really thought that we were rich folks."**

And when we finally got a refrigerator, boy we really thought that we were rich folks.

At that time they had a flat rate that you paid which was $1.50 a month for so many kilowatts, but you never went over because you couldn't afford to.

Everything that used electricity and at that time it didn't go by Horry Electric, it went by Willie Wire Hand.

Social Illumination

Electricity brought us pride

by Daisy Forrest
Bethune, South Carolina

Electric lights are a blessing from God and certainly a gift from heaven. Santee Cooper, who first made electricity available in South Carolina on Feb. 17, 1942, brought electricity into our home in 1948, I was 18 and knew it was the best thing that happened to my family in those days. Before electricity, things were chaotic. We burned a kerosene lamp for light. An avid reader since childhood, I didn't read at night. The only kerosene lamp we had didn't give enough light to read, so I only read during the day.

We cooked on a wood-burning stove. I often had to cut wood for this purpose. I felt as though this was unfair but I was the only child in the house. Grandma would cook some of her best meals hunched over that old wood stove.

My community changed when we received electricity. Everyone was proud. We all felt more safe at night by having the ability to see our surroundings more clearly. We could see who stood on our front porch. If there was a bump in the night, we could peel back our curtains to see what or who caused the stir. I could now read at night which was more convenient for me — most of my pre-electrical days were spent cooking, washing and pressing clothes and cutting wood. The benefit I appreciated most was that I no longer had to burn my hands retrieving the smoothing iron out of the wood stove. I had to press the town doctor's shirts with an all-metal iron heated by placing it in the stove. With electricity, Grandma bought me an electric iron that I guarded with my life.

Even today, I try to teach my grandchildren to appreciate what they have. However, I am one human being who will never forget the joy and pride I felt the day the lights came on.

Circle of Brightness

We caught up to city folks

by Darlene R. Hough
LaFarge, Wisconsin

The coming of rural electrification definitely changed my life during my teen years! My parents, a younger sister and I lived on a small farm in Vernon County, Wisconsin. I was a "Depression" baby and grew to my pre-teen years without electricity. This meant that we had no running water, electric lights, or indoor bathroom. We cooked on a wood range, carried in all our water, heated with wood stoves, and used the outdoor privy. It was an accepted way of life, and all our rural neighbors lived the same way.

However, as I became aware of how other people lived, I could see that our town friends and relatives had a great convenience that we didn't have — electricity. They didn't have to carry in wood and water, clean lamps, or go out into the cold and dark to the privy. They had refrigerators and toasters and their lights came on with the flick of a switch. What luxury!

My parents milked their dairy herd by hand, and as I approached my teens, my father began to strongly suggest that I should start helping with the hand milking. I was terrified of the cows, and I hoped some miracle would save me from having to get close to those big, smelly animals twice a day. A miracle did happen — rural electrification.

After World War II, rural Wisconsin became rapidly electrified. My father quickly signed on to have our farm wired, and he began plans to install all the modern conveniences. He was very willing to make life easier for all of us.

I was 12 when the lights came on. In quick succession, we not only had lights but a furnace, running water,

Social Illumination

> "Now, when I went to school, we farm kids were just as modern as the town kids because we too had an electrified home."

refrigerator, toaster, and the first bathroom in our rural neighborhood. And we had electric milking machines. I was saved from the cows! It made quite a difference in a teen-ager's social life if she didn't have to go to the barn twice a day and then have to clean off the barn smell before going to school.

Now when I went to school, we farm kids were just as modern as the town kids because we too had an electrified home. We were proud to invite our friends to our home. With our brand new appliances, our homes were often more up-to-date than the ones in town.

The drudgery of many farm jobs was dashed by electricity and my parents could enjoy more free time with my sister and me. I decided that a farm was a nice place to live after all.

After college, I taught in a one-room country school where electricity was an important aid in education. I married a farmer, and we are still on the farm — an all-electric farm, of course. We have been members of the Vernon Electric Cooperative for more than 50 years. Electricity has definitely brought many wonderful changes in my life. I still have a feeling of appreciation each time I turn the switch, and the lights come on.

Circle of Brightness

Quality of life improved

by Hazel Adams
Conway, South Carolina

I was a city girl all my life until I married a G.I. during World War II. We lived in a couple of cities until finally moving back to his country roots in a tenant house on his father's farm in November of 1948, which was located just off the highway a short distance.

My husband wired our house but due to war-time shortages and material, we didn't get electricity until some time in 1949. People out on the highway had electricity when we moved here, so I'm not sure how long they had it but it was a short time. We left our electric appliances at his cousin's house who lived on the highway. I walked over a mile to do my laundry and ironing and this city girl had to learn to cook on a wood burning stove.

I tried embroidering at night but I was so blind by lamplight that by daylight the color of thread I thought I used was not right at all. Oh how I dreamed of being able to flip a switch for lights and to be able to have my appliances with me again. To turn on a spigot and have water, and to be able to have a bathroom again and not an outside privy!

> "I tried embroidering at night but I was so blind by lamplight that by daylight the color of thread I thought I used was not right at all."

It was one of the happiest days of my life to be able to flip that switch for lights and have my electrical communications again! It was a little later before I could enjoy the luxury of a bathroom!

In my community, people farmed tobacco, Electricity meant no more hot wood furnaces to cure tobacco. They could have freezers to store their meats and vegetables. They could have electric fans instead of hand-held ones in their churches and homes. Later, they had air conditioners and electric heaters instead of wood and coal heaters if they so chose.

I have thanked my electric cooperative in my heart over the years for making things easier for me and the community and for cooperatives everywhere for adding to the quality of life for rural America in so many ways.

Electricity a blessing

by Minnie Stamey
Pickens, South Carolina

It was like walking in the dark with only an oil lamp to see by, it was hard to see how to do anything, with only a lamp. We always had to do things like sewing or crocheting or knitting in day time as it was hard to do at night. Also canning our home-grown vegetables in day time, it sure was hard going in those days.

When we got electricity, it was so new to us, we felt like we were in another world. It was just wonderful to be able to pull a string, and turn on a light. We could do things at night that we were not able to do before. I was a small child, but I will remember it forever. It was just a blessing from God, and was greatly appreciative.

> "When we got electricity, it was so new to us, we felt like we were in another world."

As I best remember, it came through our way about the summer of 1938 or 1939, but we were glad to get it and it did make a big difference,.

We didn't know hardly what to do with ourselves.

We give God the praise for sending it to us. He knew we needed it. It was just a blessing from God above.

: Social Illumination

The era of convenience

by Mary Crites
Varnville, South Carolina

We could not have running water so of course we could not have an indoor bathroom. We had to be careful and not waste our ice. In other words, we had no conveniences before electricity came along.

The day electricity came into our home it was a hot summer's day. My sister and I were very excited. We knew we would be able to keep cool with electric fans. In the fall, my sister and I could stay up after dark just to study if we needed to.

Electricity made things in our community much easier for everyone. We could all have running water and hot water tanks to heat our water to take a bath. Pumps were put down to get water, therefor our water was cleaner and more pure.

The ladies were ecstatic about getting a washing machine to wash our clothes and an electric stove which could prepare our food faster.

It was such a pleasure to see our father beam with pride because he was able to provide some conveniences and make life easier for his family.

Electricity came into our house in about 1951 — which meant no more oil lamps. We could have all the ice we needed for cold drinks with our new refrigerator.

In a few months fall would be here and of course we were anxiously waiting for the month of December so we could put electric lights on our Christmas tree. We were so excited when the month finally rolled around. We begged our mother to get a tree real early that year. She agreed.

I believe that was the happiest Christmas of my childhood.

Electric radio was greatest

by JC Penland
Ware Shoals, South Carolina

During the Depression, if it hadn't been for the catfish in Duncan Creek, we would have starved to death. We lived close to the creek which was below Reno School in Laurens County, South Carolina. My parents had seven boys and our grandparents Sam and Josie Mobley, lived down the road.

That was in 1933. One night, by brothers and I were sent to Grandmother's and when we returned home, we had a brand new baby sister — delivered by lamp light She was brought into the world by the local "granny woman." Mother named the new baby after Grandmother Mobley.

Some people had battery-set radios and listened to the "Grand Ole Opry" on Saturday nights. Some people talked on and on about listening to Roy Acuff of the "Grand Ole Opry," but the radio was just something else we couldn't afford.

> "Some people talked on and on about listening to Roy Acuff on the 'Grand Ole Opry,' but the radio was just something else we couldn't afford."

Mother had three black washpots which were kept close. My brothers and I drew water from the well to fill the pots. Mother would build a fire beneath the pot designated for boiling the wash and that fire warmed the water in the other pots. She rubbed out the dirt on a scrub board, then rinsed the wash in the other two black pots. Finally, with her cherished clothespins, Mother hung the wash out to dry. Pressing was done with fireplace heated smoothing irons.

Grandmother had an icebox and had a 300-pound block of ice delivered to her each week. It was a real treat when Grandmother made ice tea for us.

There was a cotton quota. Very little was allowed to be planted, maybe two acres. If the quota was exceeded, they'd come around and make us plow the cotton up.

Back then, we'd be in the field working, and Daddy would send Mother home to prepare supper which she cooked in a wood stove. At times, she had trouble firing up the stove, and the result was that Dad would smell the kerosene smoke, then later scold Mother for wasting it.

She'd just grin.

We'd sit on the porch after dark and wait for the evening meal. Then, with our table lit by lamplight, we'd eat cornbread and buttermilk or biscuits and catfish. After supper, we washed our feet and went to bed.

Later, we moved.

I don't recall the exact time the Co-op put electricity in our house. Ours was an older home in which wires ran across the ceiling to light bulbs hanging three or four feet from the ceiling.

There wasn't anything fancy the day the lights came on at our house. The electricity lightened our mother's workload. The biggest thing in all our lives was daddy's old radio.

Daddy finally got a radio somewhere from somebody. It was an old radio when he got it. Then, on Saturday nights we listened to the "Grand Ole Opry." There was more static than music but we thought it was the greatest thing, that radio.

Circle of Brightness

Life's joyful new dimension

by Ruby Pope
Ravenna, Nebraska

The day the lights came on is a day we'll never forget. After we got electricity it was as if we had been living in the "Dark Ages," and really we were.

Life on the farm took on a whole new dimension. Neighbors called neighbors, and everyone rejoiced.

Our electricity was turned on in January of 1950. I kept our first bill of $3.50. The electric bills are much more now but as every farmer would say, "Its worth every penny or dollar."

My husband's brother and wife lived on the same yard we did and farmed together. The first night we had electricity my sister-in-law came over. It was so light in every corner of the kitchen. She said "I just feel like I want to stay dressed up all the time." It didn't take too long before we got over that idea.

> "The electric bills are much more now but as every farmer would say, 'It's worth every penny or dollar.'"

When the men went out to milk, which they usually did after supper, you can't imagine how nice it was not to have to carry that lantern. The yard light even showed the way out to the barn. I wonder what the cows thought! Talk about the lantern, it was a morning ritual that we had to wash the lamp chimneys to have them shiny and ready for the night.

I had an old range stove at first when I got my electric stove. I baked all my bread in those days. The electric stove didn't seem to brown the bread like the old range did. I knew just how much wood or cobs to put in to get that nice brown crust. Yes I got over that foolishness too. The electric stove did a super job. I only had to get used to the change. My son was only two at the time. The burners on the stove fascinated him. He always thought it was fun to put his hand over the burner. We as parents warned him they would be hot sometimes and burn him. Curiosity overpowers little folks sometimes. He didn't believe us until one day he did it and it was hot — it blistered his little fingers all across his hand. I felt so bad about it but it only took one time to learn his lesson. No more fingers on the burners!

What joy to not have to run out to the windmill to carry in water for washing, baths or whatever. What joy to have a hot water heater, where all you needed to do was turn on the faucet and presto — hot water! No more filling the reservoir on the old range and oh, last but not least — a bathroom with a tub and stool. No more running across the yard with a flashlight. Maybe even brush a little snow off the seat in the outhouse. As time went by, we got an electric radio and oh my, a TV in the 1950s.

Life for us changed from the "dark ages" to days of light in January of 1950. What a joy.

Circle of Brightness

Electricity energized dream

by Belma Owens Lee
Lexington, South Carolina

When we were kids back in the 1940s without electricity, oh it was so tough and hard on all us poor folks. My grandpa had two big hogs he swapped for two acres of land in Cayce, South Carolina. Then we started gathering used lumber anywhere someone would offer it so grandpa could build our home. Grandmama worked so hard taking care of us kids and helping Grandpa. Mom was working as a cook in a restaurant in Columbia. We knew Grandmama was our Angel, as she still is to this day.

We used any thing we could to get light after dark: kerosene lamps, candles, a fire in the front yard. Kicking a chair leg or a table leg after dark is very painful but you had to be very careful and feel your way.

Grandpa and Grandma and Mom worked very hard getting our home built, but they made it, mainly for us kids. They made it so we could have a home and it was beautiful to us.

> **"Grandpa and Grandma and Mom worked very hard getting our home built, but they made it — mainly for us kids."**

In 1948 grandpa and some of his friends dug a deep well, and outhouse, the works. We carried water into the house for months, then mom had water pipes run into the house, Grandpa worked as a loom fixer at the Duct Mill in Columbia, South Carolina.

Grandmama never complained about anything much, she just worked. Our old home place stood for years after

Grandpa and Grandmama passed on. But Grandaddy did make life so much better for us before he left this world, he and his friends ran electric wires from the middle of the ceilings in each room.

The electric company put in a switch box with fuses. Then Grandpa bought light bulbs and pull cords.

Then he called Grandma and said "Ruth come here I have a surprise for you." He even let her pull the cord of the light to come on. Oh she was so happy she couldn't believe it. She finally had electricity!

That was the day her lights came on.

Before electricity came into our lives, it was tough when we did laundry. We boiled clothes in a black wash pot in summer and winter. I hung clothes on the line in winter, when they would freeze almost before you'd put the pins on them. You and your hands would freeze also, oh it was rough. An old coal heater stayed fired up inside for us. When we could get to it.

My mom still has an electric bill for $4.19 from those days.

I thank God for electricity and when it came to Ellison Hill at 715 Simpson Street in Cayce, South Carolina. I thank my grandparents W.H. and Ruth Bush, they worked hard as well as my mother, Haley Owens.

Chapter 13

Special Days

In schools, churches and farms, REA changed America.

Circle of Brightness

Faith rewarded

For many rural Americans, the day the lights came on will always remain one of the most memorable days ever. For some, this landmark date was made even more memorable by someone or by some quirk of fate.

Electricity affected most everything on the farm. It was more than new appliances and nifty electrical gadgets, it was the affirmation that a better way of life was finally a reality thanks to the far-reaching impact of George Norris and his work on the TVA and REA.

At ceremonies all over the country, kerosene lamps were eulogized, then often buried in mock funeral services and oil lamp obituaries were often recited.

For many, the joy of electricity was tempered or even delayed by World War II, but after the war, the REA renewed its efforts, spirits soared, and rural electrification's boom years began.

In 1944, Congress passed the Department of Agriculture Organic Act (the Pace Act) making the REA a permanent agency. This changed the old variable interest rates the REA charged borrowers and set the new rate at 2 percent. It also pushed the pay-back period from 25 to 35 years. These new regulations extended electricity to new areas unable to meet the previous economic standards.

By 1950, more than one million miles of electric lines energized more than 75 percent of the nation's farms.

At the national convention of National Rural Electric Cooperative Association in St. Louis, Norris was given a silver plaque upon which were engraved farm buildings connected with REA lines, and the dam named in his honor.

A motion picture produced by the Farm Security Administration, "The River," was shown in Norris's presence.

He watched the text of the film, the orchestral soundtrack and the images of poverty, flood destruction, dam construction, and the hydro-charged finished product. He was so moved by the film that when the lights came on at its conclusion, the senator's eyes were filled with tears. He was said to have remarked, "This is my epitaph."

Norris ran for a sixth term in 1942, and was defeated. He returned home to McCook and died in 1944.

What a World War II soldier said in a letter from the battlefield to Norris: "No eulogy could express the deep feeling of debt that I feel this country owes to you. No other man has been courageous enough to stand firm on such a course as yours with the obvious pains."

In George Norris's Words: President Franklin D. Roosevelt toured Muscle Shoals with Norris and together they watched thousands of gallons of water pouring from the Wilson Dam. The President remarked "This should be a happy day for you George." Norris was moved to tears and said, "It is, Mr. President. I see my dreams come true."

In the words of the people, these are their stories about "Special Days"....

Circle of Brightness

Radio fueled imagination

by Lloyd Roudabush
Elmira, Michigan

It was around 1938 in a small town in Pennsylvania, in the early years of my life. I was born the sixth of seven children in a log house. There was a small stream with a covered bridge, and a large maple tree close to the house. We used to climb out a window onto a porch roof and shimmy down it.

Lots of children in the neighborhood had several kerosene lamps in the house. One in each kitchen, dining room, and living room. In the living room we had a battery-operated radio. We were one of the first families to have one. Once or twice a week, our place became a gathering place for mostly kids to listened to programs like "Baby Snooks." I also liked "Just Plain Bill," and "It Pays To Be Ignorant Just Like Me." Then we turned the lights out and turned on the scary programs like: "The Squeaky Doors," "Shadow" and "Green Arrow." The scary ones were best — not that I didn't like the other ones — but someone usually got grabbed and there would be screams from all.

> "Once or twice a week, our place became a gathering place for mostly kids to listen to radio programs like 'Baby Snooks.'"

It was popular to shoot off a gun the first of the year or the Fourth of July. One year, my father used some old slugs. All he got was a small pop. After three shots he got some good shells — not knowing the three slugs had lodged in the barrel. Boy! The fourth one really went off with such a forceful bang it knocked him up against the house so hard it shook the lamp outside the house. Needless to say, he had to buy a new gun barrel.

Now for the story on electric lights. We had the house wired in the 1940s. Boy what a difference. Only thing was, we were just getting used to them when dad sold the house and bought my grandfather's. Guess what? No electricity! Several months later we had electric put in there. Still no furnace or running water. The pump was on the back porch and just a wood burning cook stove for heat

Mom died in 1948 and in 1950 or 1951 we had a furnace put in. Also we remodeled the inside of the house with running water and a bathroom.

The log house was remodeled with siding, the huge tree cut down and the covered bridge torn down and replaced with a very plain flat one.

What a difference 60 years can make.

Circle of Brightness

Mother's 51st birthday

by Eldor Bock
Pleasanton, Nebraska

I remember the day very well. It was Feb. 14, 1950 and I was 13. We lived on a farm near Poole, Nebraska in Buffalo County.

The day was my mother's 51st birthday and in those days all the neighbors knew everybody's birthday.

On the night of the birthday, they would all come and bring either sandwiches or cake and play cards until midnight. We always had four or five tables of pitch and on these tables were kerosene lamps for light, but this day was going to be different, because in the afternoon, the electricity was turned on.

That night everyone came in a happy mood. We just had to flip a switch and we had light to play cards by and a yard light that lit up the farm yard so everyone could see when they went home.

Special Days

A five-waffle salute

by Fred B. Beck
Sumter, South Carolina

It was a great day, my sister and I had been doing our school work by the Aladdin® lamp for years, and suddenly there was more light that we knew what to do with.

For years we had looked longingly at that electric waffle iron on the top shelf of the pantry, just waiting for that great day.

True to her word, Mother found the cord and took it down from the shelf. Our greatest fears were soon put to rest — it worked! Five waffles later, with a contented smile on my face, I said thank you, Black River Co-op.

I feel sure that our whole life was changed for the better on that day in December.

Radio contest winner

by Gloria Woodruff
Ulysses, Nebraska

The lights came on for my family, in the year of 1948. What a thrill, after filling lamps and lanterns with kerosene, and cleaning glass chimneys and trimming wicks, Just push up the switch and have a bright light.

The windmill was not a dependable way to bring water for washing clothes, taking baths, filling the cattle tank, or filling the drink pail in the house. It took a good breeze to make the wheel turn, but oh, how that wheel would turn when a wind storm came up. It took a very strong person to pull the handle down, to stop the wheel from turning.

The ice box caused many problems in the hot dry summer days. The ice melted so fast and the pan under the ice box was always full and made a big mess on the floor. How nice to own an electric refrigerator!

Most all our clothes had to be ironed to remove the wrinkles, no perma-press materials. The irons were heated on the top of the cook stove. To test the iron to see if it was hot enough, you moistened a finger, and touched the iron if it sizzled it was hot.

The electric iron was one of our family's prized possessions. My mother won it on a radio contest, on Bill McDonald's Morning Show on KFAB. What excitement when it arrived in the mail.

Wash day was quite a stressful day. The old Maytag® was hard to start, the

> **"The electric iron was one of our family's prized possessions — My mother won it on a radio contest . . ."**

foot pedal was hard to push down, it took six or seven tries or more — then a loud POP-POP noise came from the engine. The exhaust pipe had to be placed outside, through a window or door. The water had to be carried in with buckets and poured in a boiler to heat on a wood stove, then poured into the washing machine. Tubs were filled with cold water to rinse the clothes. The clothes were put through a roller wringer to remove the water. The clothes were dried outside on a line. It was quite a task to remove them all frozen stiff. How thankful we are for our wonderful washer and dryer all automatic.

    The day the lights came on is a day to be remembered and rejoiced and be thankful, for our much easier way of living in today's world.

Circle of Brightness

When a whole town conspired

by Julia V. Dawson
Logansport, Indiana

A woman I once met told me of a powerful memory she had about her first Christmas with electricity. She grew up in Delphi, Indiana during the Depression years at a time when most houses had no electricity.

When rural electrification became a reality, her father — who worked for the electric company — spent many hours away from home teaching men how to install electric wiring.

Returning home late for supper, he would toss a cloth bag on the table. As her mother picked up the bag, a metallic jingle of coins echoed through the room. Her mother hastily opened the bag and counted out several 50-cent pieces, the fee her father charged each man for his electrical lessons. In those days, 50 cents was a lot of money.

What the woman — and the children in Delphi — didn't know was that her father and the men he was teaching to wire homes were conspiring to hatch a neighborhood Christmas surprise.

At the appointed time on Christmas Eve — as if by magic — every house in the neighborhood blazed for the first time with the brilliant lights of electricity and Christmas.

The woman remembers gazing in awe at the aura of brightness that radiated around the head of the angel perched atop the tree. After absorbing the grandeur of seeing their own beautiful Christmas Tree, the children ran up and down the street, in and out of every home in the area, savoring each moment as they gazed at lights emitting from every window in the neighborhood.

Ice cream memories

by Martin Anderson
Guntersville, Alabama

It was September 8, 1948 in the southern Appalachian Mountains of North Alabama, in the Swearengin community of Marshall County.

For months our small farm house had been wired for REA power. We had purchased a refrigerator, churn and iron so we could begin using power immediately as well as turn the lights on — when the electricity was turned on.

I was an anxious fourth grader who couldn't wait for the lights to be turned on, and to have real cold ice cream on those hot afternoons after school. My mother promised, after my repeated requests, to have plenty of ice cream mixed and ready to freeze when the power came on. Someone told me at school that our power had been turned on that afternoon.

I ran all the way home - a mile from school and all the way to the refrigerator. I opened the door, removed the frozen tray, lifted the handle to loosen the block of ice cream and ate them one after the other with my hands — no bowl, just my fingers. Now that was exciting and especially good because we normally had homemade ice cream only on the Fourth of July.

Getting electric power in our rural community was the biggest news since World War II was over. So when the stakes of the poles were set, no one dared to touch or move them even if it was in the yard or driveway. It would have been O.K. with me and many in the community to set a pole in the chimney if that was required for electricity.

We were thrilled that we could have bright lights like the city folks around us had enjoyed for years.

Circle of Brightness

A new dress worn with pride

by Rusha M. Pringle
Ridgeville, South Carolina

When we first got electricity, I was about 12 years old. We used a lamp to cook by many times. We didn't have a shade to cover the lamp, but when the electric lights came on, Lord, this was a blessing we could not forget. I read my lessons by the lamp light and it made such a difference when the lights came on.

My mother used to sew clothing for different people and when the lights came on, she could finally see how to make a dress. I remember she made me a school closing dress, when I had to walk almost three miles down Dixie Highway, an old dirt road going to Little Rock School of Cottageville, South Carolina. My brother and I sang a song for the closing of school. He invented his own little musical instrument and our song was "Oh I Must, I Must See Jesus For Myself Some Day." Everyone enjoyed that little song so much. They were applauding so much, we felt as if we really had done something! It made us very happy. He wore his little white shirt and white pants. I wore my little white dress my mother had made for me by the electric lights. It was pretty.

> **"My mother used to sew clothing for different people and when the lights came on she could finally see how to make a dress."**

I was in the fourth grade and we only had one teacher back then in a classroom of about 15 to 20 children, but we all got our lesson. One of our teachers drove from Round O, South Carolina, Another teacher walked to school with us. We went

Special Days

to school on rainy days. Our feet got wet, frost was on the ground, and our feet had gotten very cold. They did not feel like feet. After walking to school down Dixie Highway, we had to climb through the window to make a fire in an old wood heater. We used dried pine cones to make the fire. By the time the school house got warm, our feet were still not warm. We unthawed about noon or 1 o-clock, but we lived and learned through it all.

I think this school went to the seventh grade and about the time I finished seventh grade, I had to ride the bus, another great blessing to me.

Thank God for this. Thomas Edison invented electricity. God bless him in his grave. He did a marvelous thing which we will always remember the blessing God worked through him.

This has been my memory of the day the lights came on.

Circle of Brightness

Electric radio captivated us

by Arvilla Schuman
Bad Axe, Michigan

We blew out the lamps for good sometime in the early part of 1938 when we were hooked up with Thumb Electric Coop in Ubly, Michigan.

The first electrical appliance we had was a small radio Mother purchased from the Spiegel® Catalogue. It was delivered before the juice was turned on. My sister and I hurried home from school many days and turned it on expecting it to work. Finally it worked, I was about 8. I remember laying on the couch crying as I watched my dad nervously pace the floor and check outside as we listened to "War Of The Worlds." That was the Orsen Wells broadcast in October of 1938. Other favorites were "Jack Armstrong — All American Boy" and "Ma Perkins." We never had a radio before electricity.

Also, before electricity, we used oil lamps, except on very special occasions when we used a gas lamp. We were always happy for it was much brighter. Imagine our excitement when Mother bought floor lamps with a bulb that made three degrees of light just by turning one switch. We no longer had to sit directly under the lamp to see to read. Mother no longer took us upstairs to bed with the lamp. We couldn't carry it ourselves. We could flip a switch at the bottom and turn it off at the top. What fun!

In time, wash days changed. Mother pumped water at the well, carried it to the house and heated it on top of a wood burning stove. She dumped hot water into the gas-powered machine. Sometimes the motor would start and other times she pumped away on that foot starter and nothing happened! The hot water heater came later when money and priorities allowed.

Then Mom bought the refrigerator on time credit. We no longer used the large crock, buried on the north side of the house. Milk, butter, and whatever else we could fit in it was stored there to keep cool. It was covered with a wooden lid and stone to keep animals out. When the refrigerator was bought, we made popcicles, ice cubes and ice cream. What a treat!

The water pump was powered by a gas engine. Dad started it and pumped the tank full, but on hot days when the cattle drank extra and he was working, we pumped the water by hand. Rest assured, we never overflowed the tank or pumped the well dry like dad did when he put the electric motor on the pump and forgot to shut it off on time.

Luxuries came much later, one being Christmas lights for the tree. They replaced the candles we lit in the evening after the work was done. Everyone had to sit quietly and watch the tree while they were lit as the slightest movement made the flames flicker and could cause a fire. Of all the things that have been replaced I miss those Christmas candles the most.

A freezer replaced the "locker plant' in town — a walk-in freezer, building that contained drawers people rented to store their frozen food. You took home what ever you could use or store at home. Milking machines were a real time and work saver for all farmers. They no longer had to milk by hand, or cool the milk by placing milk cans in the water tank.

On the farm, we bought things needed in the barn or outside first. Things for the house came slowly over the years. First a stove, hot water heater, and finally things like mixers, toasters, and vacuum cleaners.

Electricity changed the way people live. We take so much for granted now that its hard for people to survive when the power goes off. It happened in our town last week. The stores couldn't get into their tills. Clerks couldn't make change without their calculators, or automatic tills. Some doors wouldn't even open! This is a world run on electricity.

Circle of Brightness

A prize-winning letter

by Dorothy Pollard
Milburn, Nebraska

We had our house wired and ready for the switch to be flipped whenever Custer Public Power District said "Turn it on." In April of 1952 the "go-ahead" was given. I remember so well May 11 — a Mother's Day gift from my husband Glen and our two boys. It was a Betty Crocker® electric iron and cook book, my first electric appliance! The iron is still bright and shiny and works, the cook book is quite the worse for wear, but still very special.

In May 1953 we added a baby boy to our family, creating lots more washing and drying of diapers.

The next spring Custer Public Power sponsored a contest, "Why I'd Like to Win a Frigidaire® Electric Dryer," in 50 words or less and the prize was a dryer. I entered that contest and time went on.

Several months later, a furniture company truck drove up in front of the yard. I thought the delivery man just stopped to ask directions. Imagine how I felt when he said it was mine — A dryer! I'd completely forgotten about entering that contest!

Several friends asked what I'd written and I'd just smile. Soon, however, they knew the secret. That fall, September 21, 1954 we had a baby girl. That dryer was my prized possession and used for many years. Through the years we gradually acquired other electrical conveniences. Anyone who grew up without electricity and hot running water doesn't ever take these modern conveniences for granted.

I continue to count the many blessings through the years and 1952 was among the most blessed year.

Country living improved

by Leonora C. Burbage
Summerville, South Carolina

It was probably the early 1940s when electricity was delivered to my home. It was an exciting experience — magical. When the light bulb lit up, Edison couldn't have been more enthusiastic than me; it shed a whole new "light" in country living.

Our little community church held services only morning and afternoon unless a revival was held at night. If this was the case, the lamps hanging on the walls had to be lit and the pot-bellied stove had to be fired up. During the warm summer months, we were kept cool by the use of complimentary fans from the local funeral home. Now, we have a large church with electricity and can have services anytime, day or night, hot or cold. We can also enjoy "dinner on the ground" with the modern conveniences of stoves, ovens and microwaves.

The larger power companies would not serve communities in rural areas because no profits were expected to be made. Electric cooperatives saw a great need and fulfilled that need for power to all rural areas, and I thank them.

Electricity has brought so many modern conveniences into the rural home; words cannot express in detail the wonderful things we now enjoy!

> At night we turn off the light
>   and go to bed to rest
> Now, this is country living,
>   finally, at the very, very best!

'Sparks' from Halloween fire

by Donna L. Calvin
Oconto, Nebraska

In the fall of 1951, the rural electric lights came on in the Montrose country school in Custer County, Nebraska, where I taught. I was 17.

When I started teaching in the Montrose school on east Redfern Table, September of 1951, the only lights we had were around 15 windows in the large one-room school. We had a small room for books and another small room for coats, a water crock jar and a wash basin on a table. There was an entryway between the two small rooms that went onto a small wooden porch.

When we had a Halloween Program the evening of October 31st, we used about six gas lanterns hanging from a wire loop which was attached to the ceiling of the school house. In addition, we had four kerosene lamps, setting on the piano top, teacher's desk and a bookcase for lights.

One of the fireproof Jack O'Lanterns caught on fire that Halloween night. I ran out to throw the Jack O' Lantern that was on fire off the wood porch. I ran back to get the fire extinguisher but ran into a young man who became my husband of 44 years. An older man behind this young man went for the fire extinguisher and put the fire out.

> "I ran back to get the fire extinguisher but ran into a young man who became my husband of 44 years."

By the time we were ready for our Thanksgiving

program the third week of November, we had electrical lights. No more lighting gas-lanterns and having to pump the lanterns up with air every little while for the lights to stay bright. The kerosene lamps had to be watched, so the fire on the wicks wouldn't smoke up the glass chimneys if the flame went to high. Also, no more (supposed to be) fireproof Jack O-Lanterns with candles in them on the school's wooden porch which lighted the entryway.

It was a delight and joy at the Thanksgiving program to turn the switch on for the electrical lights without wondering if any unexpected eventful fire would happen.

No more use of smelly kerosene lamps and pumping air into the gas lanterns to make the lights brighter which the parents had brought from their homes. With electrical lights, the lights always stayed on, evenly. No brighter, no dimmer.

Circle of Brightness

Electric Christmas glow

by Judy Calmus
Howard, South Dakota

Christmas . . . smells and sounds stir our memories. Lights . . . ah. Lights.

My husband says the sight of the Christmas lights in his rural South Dakota home stand out when he recalls the difference rural electrification made in his young life.

What a difference a day makes? What a difference electricity makes!

Growing up on the outskirts of a small South Dakota community, my favorite Christmas memory (of which I often draw upon) is the nostalgic, 12-block walk to midnight mass my brothers and sisters and I non-complainingly, but excitedly took with our mother.

Walking in the street due to the snow-covered sidewalks — I can, if I close my eyes and open my memory chest — see and freshly feel the wonder of the large fluffy snow flakes, so brilliantly adorned in a blue hazy brilliance of the electric street light . . . leading us to the Christ child, as the star once led the Magi 2000 years before.

As I raised my family in rural South Dakota, I have never taken electricity for granted. Sometimes, due to weather-related outages, we've been cold; I've thought, "I'll keep warm by ironing" — no I can't. "Well perhaps if I bake" — no, I can't. "Well there's always TV" — well no there isn't.

Electricity — not just memories — it's life.

Thank you God for giving man the knowledge to bring us light and REA for bringing it to us

Special Days

Baby couldn't wait for lights

by Mrs. Leone McLaren
Millington, Michigan

Here in mid-Michigan before 1938, if you lived more than a mile from the main road your hopes of getting electricity to your farm or home were nil!

An uncle lived a mile west of Fostoria in the early 1930s and Edison came to their farm because they agreed to buy two appliances and a certain number of their neighbors had to make the same agreement. The new refrigerator was put to use but the stove was kept as a polished jewel to be admired!

Early in 1938 we moved into an old farm house on Barnes Road, Watertown Township, Tucsola County. That spring we scraped up enough money to get the house wired; very sparsely by today's standards but quite wonderful to us then. Rural electrification was coming!

I was due to deliver my second child the middle of June, at home of course. He wouldn't wait and was born by lamplight at 4 a.m. on June 11. Exactly two weeks later the power was turned on. That date is etched in my memory though it would be years before we could afford indoor plumbing and other luxuries.

In 1966 we had to move to town for four months due to a house fire. With that exception, I have paid my bill to R.E.A. Power for more than 60 years.

Thank you for being there for us.

Circle of Brightness

A Christmas Eve miracle

by Beth E. (McLeerey) Graham
Lyons, Nebraska

It was December of 1938, when we students of the Parks High School, east of Rosalie, began to generate a certain amount of expectancy. Every few days a bit more work was in progress to bring electricity to our school and the surrounding farms.

Parks School was a 10-grade school system that boasted two large rooms on the upper level. The lower level contained a fair-sized class-room, a furnace room with two coal-burning furnaces, a coal bin, a cob room, and a small play area.

It was among the largest high schools in Eastern Nebraska, if not the largest. The rooms on the upper level each had a long closet. There were windows in the closets which were of great help, but the rooms were then reduced to having windows on only one side.

On sunny days the room on the east had plenty of light in the mornings while the west room had the sun in the afternoons. Cloudy days made reading and writing much slower and tiring. Only occasionally were the kerosene lamps, hanging by the chains from the ceilings, lit. They were made of brass with the bowl for the kerosene shaped in quite a large circle. The glass globes or chimneys were tall with a slightly reflective shade that fit about a quarter of the way down to the globe.

The wicks were rather difficult to trim so that the flame would burn in a smooth half-circle. Any points on the wick meant an uneven flame prone to smoking the chimney black.

Regardless of a clear or smoky chimney, it seemed my desk was always directly under a lamp, which meant the bowl of the lamp cast a lovely shadow on any work I might try to accomplish. That was also double trouble on a dark and rainy day. No wonder we were exuberant just think even lighting no matter where we sat. Also, no smoky chimneys, no smell of kerosene, no fear of fire and the whole room full of light at the click of a switch!

Finally, word came that the installations would all be finished in time for our Christmas Program. Now Christmas Programs were always special, but this one had the most suspense. Everyone worked hard at memorizing, decorating and even cooperating to keep the noise level down.

The old lamps were taken down and set aside. Everyone came by foot, car, or wagon quite early. We could hardly contain ourselves But there were no lights because for some reason, the current wasn't coming through.

The REA team worked diligently. Time for the program to start came and went — finally, the old lamps were readied and hung above the stage. Sure enough, the shadows of the lamps were going to shade some very important areas, like the tiny tots Christmas acrostic, some decorations and pertinent costume arrangements.

We flipped switches several times but, still no lights.

No one knew if the switches had been flipped off or not, and at the moment, didn't really care. Our expectations had fallen flat!

So it was decided to start the program, since it was over an hour late.

Just as everyone in the audience found their seat and silence behind the curtain had been accomplished — "Ohs" and "Ahs" filled the room with one accord.

Several switches had been left on and our lamps without shadows had suddenly filled the room with unbelievable brightness.

The REA team became the heroes of the night.

Ah, yes — the Christmas when the lights came on!

Chapter 14

Yesterdays Today

Norris's legacy is his impact on a rural countryside.

Norris's legacy

Rural electrification and the Tennessee Valley Authority are just a small part of the George Norris legacy.

Among his other noteworthy accomplishments were the overthrow of "Czar Cannon"; his support of Al Smith for President in 1928; his filibuster against the Armed Ship Bill in 1917; and his opposition to America's entrance into World War I.

Thirteen years after his death, a special Senate committee decided to honor the five most outstanding Senators by hanging portraits in the Senate Reception Room. Senators compiled a list and so did a panel of more than 100 scholars from the nation's top universities. George Norris topped the scholar's list, but when the Senate revealed its top choices, Norris was not there. Subsequent studies have determined that contemporary senators like Norris may have been prejudiced on the Senate list because their political opponents were still active.

Richard Lowitt, who wrote several books about Norris's life and his concern with the common man, said, "It was this concern for people as victims of circumstances over which they had no control, be it corrupt or intensely partisan politicians or overzealous and inhumane corporations, that led Norris to modify his views and begin to move along political paths in ways that would leave a marked imprint on the American scene."

In George Norris's Words: "There was that warm spring afternoon when mother, who had been busy throughout the entire day, called to me to assist her in planting a tree. She had dug a hole, and she wanted me to hold the seedling upright while she shoveled the dirt in around its roots and packed it tightly. I looked up at her, and it came to me she was tired. The warmth of the afternoon and her exertions had brought small beads of perspiration to her brow. So I said to her.

"Why do you work so hard mother? We now have more fruit than we can possibly use. You will be dead long before this tree comes into bearing: The little farm was well stocked with fruit. It had its apples, its peaches, and is sour cherries.

"Her answer was slow to come, apparently while she measured her words.

"'I may never see this tree in bearing, but somebody will.'"

In the words of the people, here are their stories of "Yesterdays Today"....

A modern-day 'pioneer wife'

by Billie Gilbert
Livingston, Montana

It had been our dream for many years to retire and build a new home on the old home place. As usual, when you have your heart set on something, you ignore everyone's advice and warnings.

The weather had been unusually cooperative for January so we started the foundation in February. From then on the process drug on and on and on. In the spring it started to rain and didn't stop until summer, one of the hottest I have ever experienced in Montana. However, we were still happy and optimistic, but made slow progress. Fall came and we weren't even close to a move-in date.

It was time for Plan B. My husband says, "Well we could stay in grandpa's little house until ours is ready." I resisted, but it sounded all right and we could save rent money. There was just one tiny problem — although the little house was insulated well and had a propane heating stove, we wouldn't freeze, but there was no running water, no bathroom, no phone — and no electricity.

Being a city girl all my life, I was willing to take up the challenge of "pioneer life." We moved our essentials on October 31 — Halloween. How appropriate for what was to come.

The first night it was dark at 6:30 p.m. What to do? We couldn't watch TV, no electricity, and the oil lamp made it impossible to read. So we went to bed. The next morning I used a propane, two-eye hot plate to cook breakfast. That was bad enough, but now to wash dishes. Now I walk through snow, wind and cold to the spring to get water. No electric pump, so I heat water on the stove. With no electric hot-water heater, I wash dishes by hand. Everything takes longer the old way. I am ready to be a city girl again.

2 a.m. should be sleep time, but nature is calling and out I go in the dark to the house out back. Remember — no electricity. So what do I do? I run into the barbed wire fence in the dark. No stitches, but my respect for modern conveniences is growing.

My next challenge is washing clothes. I am not going to break the ice on the pond, so it is a five-hour trip into town and back because there is no electricity. Lets see, I miss my electric toaster, my electric curling iron, my microwave, my oven and on yes, my coffee maker.

> **"I will walk to the front door, turn the key, reach for the light switch and my REA electricity will make my life bright and easier — again."**

I am thankful to our ancestors for the hardships they endured and the sacrifices they made, but please don't tell me how much I will appreciate the new house when we move in. I know! I can see it now. I will walk up to the front door, turn the key and reach for the light switch and my REA electricity will make my life bright and easier — again.

My story of rural electrification must end here. It is getting dark again and I have — no electricity

Honored posthumously

by Virginia Spelbring
Poland, Indiana

June 17, 1948 — a day I'll remember as long as I remember anything. I took a snapshot of the transformer being installed on the pole in front of our house. When I went in the house, the refrigerator that had been there for several months, was running. I opened the door and the light came on. Magic!

I had been heating laundry water in the basement on a circa-1900 range and used a gasoline wringer washer, with the exhaust hose out an open window — even in winter. It made so much noise I couldn't hear our one and three-year-old children upstairs. No more sad irons or kerosene lamps to fill and clean daily, radio music filled the air and a bathroom was possible. Soon a freezer lessened the canning.

Farm chores changed too. No pumping water to carry. Kerosene lanterns were retired and later grain dryers, welders, heat lamps for pigs, automatic waterers and then computers for record keeping were possible.

Two or three years before we were married in March of 1942, my husband Harold, along with men from three adjoining counties had organized to secure electricity for this rural area. Because he was the only single man in the group he signed the right-of-way easements. Then Pearl Harbor stopped the plans. When the war was over, the directors were told this was too small an area to be feasible so recommended we be divided to become members of three adjoining Rural Electric Membership Corporations (REMC). All this time we could see the lights of a public utility just a mile away.

In February of 1948 our neighbors along the road got power. Our house is 1/3 of a mile off the road and because a different sized wire was used for that distance (and was unavailable because of war time shortages) we weren't connected until June 17. It did seem ironic that my husband Harold had driven many miles attending meetings, signed the easements and we were the last to get power — four months after our neighbors. However, now we know his work all those years ago finally paid off.

After his death in 1988 from complications of Alzheimer's, I had reason to relate the above information to our REMC Office. The General Manager drove over 50 miles to our home and presented me a gold-lettered certificate. It named Harold a "Pioneer in the Rural Electric Movement.

> "The (REMC) General Manager drove over 50 miles to our home and presented me a gold-lettered certificate."

That certificate hangs where I see it every day.

Where was I when the lights came on? Right here where I still live. Oh the changes I have seen. My cordless phones won't even work if the power is off. Thank God for all the people who worked so hard to "light up the country" and prove we are not the hicks and hayseeds we were (and still are according to some).

All we needed was a chance and power provided it.

Circle of Brightness

Old memories, modern light

by Gene Erickson
Arcadia, Nebraska

I sit here by my small table in my small but comfortable enough house in northern Sherman County, Nebraska, just four miles south of the Valley County town of Arcadia. As I contemplate what would be appropriate to write about for an essay on "The Day The Lights Came On," I am very pleased to have at least 100 watts of light shining on my work.

I am also very pleased to hear the electronically controlled furnace come on to keep me comfortably warm. I am pleased to hear the automatically controlled jet water pump come on to provide me with water to wash my hands and face and my dishes. It also looks nice to see the yard light come on in response to the electric eye.

Then, my thoughts go back to the middle 1930s when I used to bring some school books home from Arcadia High School, where I was a student at the time, and study them by the light of a kerosene lamp. Or when I would light the old kerosene lantern and trudge down to the barn with the milk pails and lantern and milk seven or eight cows by hand.

> **"'A fellow almost has to light a match to find the light in this place.'"**

I think of the time that the "Farmer's Holiday Organization" held a meeting in the country school house which was near our home. The school house was lighted that night by a small kerosene lamp with a poorly trimmed wick and a badly smoked up chimney. One farmer, who arrived at the meeting remarked, "A fellow almost has to light a match to find the light around this place.

I also think of how we put new mantles on the gas lamp. Sometimes we'd fill a new generator with high-test gasoline to gain better light for reading, only to have a miller or two fly through the mantles.

Also, I want to relate a story that my sister, Ruth, shared with me not long ago. Ruth was teaching at the Bristol School (also known as the Cole Creek School) which was located in Washington Township of Sherman County, about eight miles sough of Arcadia. This was the school term of 1940-1941. They closed school for at least a week for Christmas Vacation, and was to open again right after New Year's Day in 1941. The weather had turned very cold during the last days of 1940 and the first days of 1941. My Dad took Ruth in a high-wheeled wagon and a team of horses on the afternoon before school was to open the next morning. After spending the night with a neighbor, she walked over to the school house a short distance the next morning and when she entered the school to start the fire in the coal and wood fired stove, the temperature in the school stood at exactly 30 degrees below zero.

The school had been decorated at Christmas with red and green crepe paper streamers across the ceiling. They gathered enough moisture in the extreme cold that they were hanging nearly to the floor, but she got the fire going and the room warmed up, the streamers retracted and pulled back up to normal position. Most all of the school children arrived, mostly on foot, some on horseback, and a normal day of school was held.

Circle of Brightness

Electricity's lasting impact

by Roye D. Lindsay
Fredericksburg, Texas

The scraps of aluminum wiring were still scattered around the house and in the new grass that spring of 1949. The snow drifts of the past winters' blizzards were gone and the birds were singing. The local electrical contractor finished his work a few days earlier. We eagerly awaited the REA line to be run into our farm.

I was 12 or 13, in the final year of my eight years at the one-room brick school, Sioux Creek; District 4 in Loop County, Nebraska. When I came home from school that memorable day, the crew was just setting the meter pole. Dad had them mount a regular hooded yard light on that pole. Would it really be possible for us to flip a switch and light our whole yard from the house? I was soon to find out. The lineman came to the door and said it's ready to go. Mom or dad probably flipped the switch next to the yard light and the beauty of that single bulb, round glass, and ceiling fixtures coming to life was true delight.

That old house near the highway in the beautiful North Loop Valley is still the home of my brother. Our great-grandfather built it in 1887. During the ensuing 112 years that $190 house has seen births, wars, Christmases, deaths and such, but flipping that switch back in 1949 perhaps affected my life and the history of our family more than any single event.

Just yesterday, the Central Texas Electric Coop set my feeder line and meter pole for my new home on the Pedernales River in Texas. We live 1,000 miles south of that old switch in Nebraska. We've all come a long way in 50 years. But we couldn't have travelled so far so fast, so comfortably without Rural Electricity!

Dad's rare story

by Rhonda Still (as told by her father, Faust Still)
Blackville, South Carolina

Faust Still has lived on Holnan's Bridge Road in Blackville, S.C. virtually all of my life.

I had just replaced a ballast in their kitchen and he recalled how they got electricity in their house in 1940.

His parents, three brothers and a sister turned on every light in their house and on their front and back porches; got into their maroon 1940 Chevrolet and rode down the road. Then Granddaddy turned around and drove slowly past their house just to witness this miraculous event. Daddy said that it was so amazing — it literally took his breath away — he was 16 at the time.

To be able to walk into a room, flip a switch and be able to see everything in the room — no words could express his astonishment!

This story means a lot to me because my dad doesn't expend too much of his energy in small talk anymore, due to cancer and removal of his larynx five years ago. But he did pick up his speaker and share this story with Momma and me.

It brought joy and a smile to our faces, seeing Daddy reminiscing and smiling about the night the lights came on in his home all those years ago.

This isn't as entertaining as the joke I heard about the woman who was fascinated about being able to walk into the room and turn on the lights — so she could see how to light her candles.

But it was thrilling for me to hear Daddy tell this simple, true story.

Circle of Brightness

A blizzard of memories

by **Ruby Finley Edwards**
Elsberry, Missouri

Today our electricity was off four hours, and it seemed an eternity! Our whole household was paralyzed — no heat, no way to cook, no water, no television, etc. This caused me to stop and remember the day in 1947 when the lights were first turned on in our rural farm home.

In the 1930s, the local I.O.U. built lines up Highway 61 — not a quarter of a mile from our house. Several times my father approached the company offering to pay to have them cross our 40-acre field to bring us electricity. They serviced only the houses built on Highway 61, and refused to build the extra line. When Cuivre River Electric Cooperative came into our area asking for a $5 membership fee it was truly a dream-come-true for our family. They built electric lines throughout our rural area literally lighting up our lives. Were it not for R.E.A. we would still be living in darkness.

My sister, Anna and I, were standing in our living room — usually a rather dark room — when the electric was first turned on in our house. We were surprised to see all the cobwebs near the ceiling in the corners of the room. We quickly swept them away, then stood in amazement at our lovely bright room. We could easily read or do our homework in the evenings.

In the past we all huddled round the coal-oil lamps in order to read. Each morning the lamp chimneys were cleaned and shined, then wicks trimmed. We always carried a lamp upstairs at bedtime, and the last one in bed had to blow out the light then climb into bed in the dark. I always tried to be the first one in bed.

The first appliances purchased were a new refrigerator

and deep freezer. How nice it was to have ice cubes for a refreshing drink on a hot day. And there was no need to hang our butter and cream in the well.

How delighted mother was to have the new electric wringer washing machine and iron for laundry. No more use for the old wash board and sad irons used to iron the many starched dresses for six daughters in school and church on Sunday — plus dad's starched shirts. This was truly a help in the work load. Of course, we learned to help with the laundry at a young age. I remember ironing Papa's handkerchiefs and our pillow slips when I was about 10, and how difficult it was to use the hot sad iron to get all the wrinkles out of the cloth and not burn my fingers in doing so.

Then my father purchased electric fans. What a relief during hot weather. Mother could iron with her new electric iron while standing in the cool breeze of the fan.

Next, we enjoyed our new bathroom. How nice to bathe in our snow white tub instead of the galvanized tub carried into the kitchen at bath time. Next to go was our wood cook stove. What a relief during hot summer days! Mother had always made beautiful angel food cakes and lovely meringue pies in the wood stove oven, but she found it much easier to set the temperature gauge on her new electric cook stove than to stoke the wood cook stove just right. A few years later gas space heaters with blowers replaced the wood heating stoves making our winters much more comfortable. Then several years after that air-conditioning was installed.

Today, we are having a blizzard. The snow is clinging heavily to trees and electric lines — breaking some. Power was off four hours, and we felt totally incapacitated. How thankful we are to have our lights on again! We are very grateful to linemen working out in the cold. Hopefully we will never return to the olden days. One outage makes us realize how truly lucky we are, and how grateful we are for rural electrification.

Circle of Brightness

The lights came on in 1994

by *Carrie L. Collier*
Ridgeville, South Carolina

In 1994, my husband was discharged from the military. We decided that we would move back to our home state of South Carolina from the state of Virginia. With the move, we found out that the state of South Carolina had certain qualifications that had to be met.

First, the land and soil had to be tested to see whether it would pass to install a septic system. After finding out that the land had passed, we could then schedule a date to have the mobile home moved onto the property. Once the home was moved, there were more qualifications that had to be met. For example, the purchasing of a pole and wiring to run wires from the mobile home to the pole. Then the installation of a septic system to be met and inspected by DHEC, Environmental Health regulations.

In the meantime, without electricity, the mobile home was completely dark. The only work that we could do had to be done during the daylight hours. When night came, we could not arrange the furnishings or work inside the home. So, we purchased a pole that was partially ready and came with some of the wiring already completed. We received a breaker case from the Berkeley Electric Cooperative, Inc. Then we had to call an electrician to run the wire from the breaker inside the mobile home to the outside pole. This work was left uncovered to be inspected by a Berkeley Electric Cooperative representative.

> "In the meantime, without electricity, the mobile home was completely dark."

Next, we called a contractor to install a septic system. After it was installed, we called the Berkeley Electric Cooperative men to come out. They brought out two poles and wiring. They ran wire from a neighboring pole to one of the poles they brought out , which was placed near the road. From that pole, they ran wire to the second pole near the mobile home. From the second pole, they ran wire to the pole we had purchased with the breaker. Then the electrical power was turned on. Thank God! What a difference the lights made. It was much more convenient to do the daily chores, (cooking, cleaning, washing, reading, etc) There is nothing else that can be compared to it. We could see clearly because of the instant lighting. With the lights on, we felt we were getting back to normal.

As for the community, they had lights and we were like the new kids on the block. The community was lit up and where our home was, it was in darkness until we received our lights. Electrical lighting makes for safer surroundings. It's easier to see from one neighbor's house to another. With lights, it's harder for someone to find a hiding place or even to try and break into someone's home or business without being seen. The electrical lighting helps make a community.

Thinking back to when there was no electrical lighting, it's hard to imagine what life must have been like in the past. Imaging how our forefathers accomplished a job or a task at night. How was major surgery was done? How much lamp oil was used? How many lamps were needed to operate a household.

Now that we have lights, electricity, we are so blessed. Thank God for our forefathers, namely Thomas Edison, for his experiments, inventions, and perseverance to achieve and to make this world a better place in which to live. Thank God for his blessings!

Circle of Brightness

An anniversary assessment

by Everett H. and Alberta M. Breyer
O'Neill, Nebraska

    Lights, lights and more lights — how well we remember that wonderful day in February, 1952 when Niobrara Valley Electric Membership energized our lives.

    In June of 1998, our four children hosted a 50th Wedding Anniversary Celebration for us and they all agreed that REA was the best thing that ever happened during our long farming career.

    We were married in June, 1948 and moved to a little farm south of the Twin Buttes in Boyd County. It had no electricity, no water system, no refrigeration, no radio and a wood-burning furnace. We were happy in spite of our simple life and we certainly learned to appreciate the little things. We'd been left out when the first members got REA. Some paperwork didn't get filled out correctly, we were told. But never fear — the second phase of lines were coming soon. It was difficult to wait.

    Our little house had to be wired and we worked hard helping a self-taught electrician finish the wiring — bare bulbs and a few plug-ins but we determined we'd be ready. We'd installed a pressure tank and pump and a bathroom (cold water only) We even had a washing machine motor and an electric iron waiting. Finally, the big day arrived.

    Since that day we've added all kinds of appliances and even built two new houses but nothing can compare to the thrill of that day in February, 1952, when our lights came on.

Old refrigerator still runs

by Ruth Huneke
Osgood, Indiana

We were anxiously waiting for electric power. Two neighbors between us wouldn't sign up and we had to have so many neighbors sign up per each mile. After several years waiting, we got enough to sign.

We hired someone to do the wiring. The big day arrived. I believe it was 1947.

When the lights were turned on, our five year old daughter ran from house to barn to see which lights were burning so she could turn them off for the first time.

We were milking cows by hand. We soon put in a milking system and went to Grade A — getting a better price.

We got an electric washer and dryer. Things were hard to buy. A neighbor told us about a store that had a refrigerator on hand. So we were able to buy a refrigerator. It had a freezer across the top. What a treat to have this convenience.

And would you believe this refrigerator is still running! It is our second refrigerator. It is very handy in the summer when we have an overflow of vegetables.

We think of the day the lights came on.

Circle of Brightness

Grandpa speaks of electricity

by Susan Medler
Morley, Michigan

I'm not sure what I expected my grandpa to say when I asked him what life was like before electricity but his off the cuff answer made me laugh.

"It was dark," he said with a slight smile and a quiet chuckle.

I knew it would be a fun thing to do with my Grandpa Tronsen who is now in a nursing home. Sometimes it's hard to know what to say and this seemed perfect for us. I would learn more about his past and he could do one of the few things left he enjoys — visiting with family

Grandpa wasn't sure what year his folks got electricity but he knew it as after he graduated from high school in 1934 and before 1939.

Electricity made many areas of his life easier. Before electricity they had to use lamps with chimneys as a source of light. The chimney had to be washed everyday. Lanterns were used to go to the barn. The lanterns and lamps didn't always work as they should. Getting water to the livestock was no quick and easy task either. They had a windmill and a gasoline engine to help get water but grandpa remembers how much trouble his dad had starting that pump.

It was up to each homeowner to get his home wired and then Consumers Power would come and actually hook them up. Grandpa's dad hired someone to do the wiring.

One of the first recollections he shared with me was when his folks got their first refrigerator. In 1939 they bought their first refrigerator in Grand Rapids, Michigan. Some people told them it wouldn't be any good because it was made during war time. However, according to Grandpa, it lasted a good many years.

Grandpa grew up in Amble, Michigan. Not all of their neighbors got electricity at the same time. Cost was a factor for some but for others it could have been the reluctance to try something new. An electric bill was also a new monthly expense. Some families just had one light hanging from the ceiling and it wasn't very safe.

> **"Grandpa viewed electricity as a great thing that made life easier and allowed them to have some frills — like a mixer."**

Grandpa viewed electricity as a great thing that made life easier and allowed them to have some frills, like a mixer. Indoor plumbing came later for them but boy did they like it. His folks turned the parlor into the bathroom. Keeping food cold was now easier because they didn't have to put up ice for the summer.

I know that electricity didn't change the every day lifestyles of people overnight, it took time and money. It affected people in different ways and to different degrees. I have always had the comforts of electricity until a storm knocks it out . . . it is moments like that, that even I can appreciate its value.

Circle of Brightness

Electricity's timely dividends

by Robert L. Cox
Holland, Ohio

I was born Jan. 20, 1921 at Williamsburg, Michigan. At the home that was my birthplace, there were several interesting features including hand-hewn eve spouts. These were the trunks of trees that were hollowed out by hand — setting underneath these is a rain barrel. From this we got water to wash clothes and for personal use. Our drinking water came from a spring at the foot of a big hill about 1/8 of a mile from the house. In the fall, boards around the base of the house covered horse manure and were staked each fall to insulate a root cellar that was under that side of the home. Each spring the boards were removed and the manure spread on the garden. The wood-shingles that covered the roof were also hand-made.

The night I was born a doctor drove a horse and cutter from Traverse City, a distance of 17 miles. He stayed at this home all night returning to Traverse City the next day. His fee was $50.

I remember my grade school years, I studied by the light of kerosene lamps. The wicks on these had to be trimmed several times each week so that the flame would burn straight and would not smoke up the glass chimney. The next innovation was the Aladdin lamp. These lights still used kerosene as fuel but had a mantle and was much brighter. Next came carbide lights. These lamps were mounted on the side of the wall and the fuel came from carbide crystals that were placed in a large tank in the ground and then activated. The light they gave was not much brighter than the kerosene lamps, but the lights were lit by a flint and much handier to use, as a match was not necessary to light them.

We had pot in those days also — one that my

grandmother cooked chicken or a pot roast on the wood burning range in the kitchen. Another pot, porcelain, was placed under the upstairs bed so we didn't have to got outdoors to use the toilet in weather that had many nights below zero.

I remember waking up different mornings with a dusting of snow on the bed covers that had blown into the room from the loosely fit windows or under the eves.

Rural Electrification — what a blessing. In 1929, my grandfather lost all his savings when the banks in our area failed. He then put his savings and mine in glass Mason jars and buried them in the ground. I believe it was in 1933, when the power dam in Elk Rapids Michigan issued new stock. My grandfather invested $2,000 with the company and thought I should put $1,000 (that I had saved by playing the trap drums in my mother's dance band) into this stock also. They were to pay dividends of four percent each year. They paid for the first year only and from 1934-1943 we received nothing and thought our stock to be worthless.

In 1943, my wife was with me with our first son at Altus Air Force Base in Altus, Oklahoma. My take-home pay was $120 per month of which $60 went to pay for our apartment. I believe it was this same year that rural electrification came to Northern Michigan. That electric company bought out the power dam at Elk Rapids paying back all dividends with interest to the stockholders.

My wife and I received a check from them about three days before Christmas. Rural Electrification had made our Christmas unbelievable. Later on we sold our stock for more than we paid for it.

We still own voting stock in that company.

Circle of Brightness

Understanding Daddy's saying

by Libby S. Lancaster
Enoree, South Carolina

Well I wasn't here yet, the day the lights came on, since I'm only about a half a century old. I once asked my daddy, J.C. (Chuck) Senn, if he remembered when power first came to our rural farming community in South Carolina.

"Yes," he said. "I was just a young boy, but I remember when they put up the first poles down the road. It is called Highway 92 now, but back in the early part of the century it was knows as Heads Ford Road. We were in the fields working as most people did back then, but being a young boy, I wanted to check out the new poles the men were putting up down the road. I wandered over and began to swing on a guide wire they had placed on one of the poles, the power line had not been strung yet. I was having a good time when Aunt Meezer came running yelling 'get off that boy, you are gonna get electrocuted!'"

> "When the sun goes down, a poor man's luck runs out."

"Yes, time was hard back then, you used oil lamps and went to bed early. I remember the ice man bringing us ice around to our homes. It would be so good and we would keep it wrapped in clean clothes to keep it from melting.

If it was hot, you slept with the windows and doors open. If it was cold, you chinked the stove full of coal or wood and slept under as many quilts as you could."

I think I know now where my daddy got his little saying of "When the sun goes down, a poor man's luck runs out."

Property completed circuit

by William Eaton
Sullivan, Indiana

I was 10 and living on a non-electrified farm only three miles from where I live now.

Before electrical power, we had kerosene lights and no inside plumbing. I remember taking a bath in a galvanized wash tub in water that mom heated outside on a wood fire.

When electric power came, we bought a pump for the well and dug ditches for water lines and put electric lights and a bathroom in the house.

I was only 10 when the lights came on. Do I remember it? You bet I do. That will stand out in my mind forever.

We sold the non-electrified farm I grew up on in 1977. We sold it to Hoosier Electric to build a power plant on.

Seems ironic that I remember that day around 1948 when Rural Electric Membership Corporation (REMC) came through with electricity to the farm. Now that same non-electrified property is used to supply electric power to thousands of people.

> **"Now that same non-electrified property is used to supply electrical power to thousands of people."**

Circle of Brightness

Eases Parkinson's effects

by Arlyce LaFollette
Lyons, Colorado

I was born in 1945, so I am probably among the younger people that learned to live without electricity.

I must share a story about an old couple who lived far out in a rural area. They were the last ones to get power to their farm. All of the neighbors and friends came for a big party when they threw the switch. They waited until it got a little darker than usual and the man told his wife to do it. As the whole place lit up, everybody cheered and clapped their hands. The old man proceeded to light the lamps and when he got them lit he told his wife she could turn it off and said. "That sure is gonna make it a lot easier to light these lamps."

I grew up on a dry land sharecropper farm in South Dakota. We were the only people in the world (I thought) without electricity. We got a weekly newspaper and dad listened to the news, market prices, etc. on the radio.

We had kerosene lanterns in the barn and lamps in the house. I always had to keep the wicks trimmed and the chimneys clean as well as filling them. We heated with a pot-bellied stove in the middle of the main room. I carried in cobs and coal. Mom washed clothes with a green speckled wringer washer that sounded like a motorcycle when she started it up. We had clothes lines everywhere. In winter, I carried in big chunks of snow to melt and heat on the kerosene stove, it took a lot of snow to fill those big washtubs but there was plenty of snow five months of the year. I remember learning how to put the clothes into the wringers without popping off the buttons, I hung clothes out and folded them and brought them in and ironed with these heavy old irons. You would keep two on the stove and clamp the first one on and use it until it was too cool then you would switch them. That was before the days of polyester and permanent press. I remember when we got a gas iron, such an improvement, the day mom got a propane cook stove too.

I will never forget the day we moved to a farm with power! The first thing dad did was buy a refrigerator and then a TV. I was so thrilled to watch American Bandstand and hear the latest hits and watch them dance. Then we got a toaster and mom's homemade bread wouldn't fit so we bought a loaf of bread. I think us kids ate that whole loaf in about 15 minutes. I still like toast but now I have an electric breadmaker and the loaves fit the toaster just fine.

> "Today, I have Parkinson's disease, and if it weren't for electricity, I couldn't be at home."

Today, I have Parkinson's disease, and if it weren't for electricity I couldn't be at home. I wouldn't be able to do the household duties.

I use a mixer, knife, food processor, can opener, and even use my dishwasher to wash the window panels from my basement windows. I rely on electricity from the time the clock radio goes off in morning until I brush my teeth with my electric toothbrush at night. The electric blanket even warms my bed for me. This letter is being typed on an electric typewriter.

I'm thankful for George W. Norris and his part in rural electrification. What a blessing it is.

Circle of Brightness

Words inadequate

by Phyllis Balzer
Kendallville, Indiana

"Please, Gram, please. Will you light the old oil lamps for me tonight?" my small grandson sometimes begs.

Scratching a match, I light the wicks and a heavy odor — which I enjoy — permeates the house. We make shadow figures on the wall, my grandson and I, and my mind drifts back . . . .

It is a zero-degree day. I am on our back porch, with several lamps beside I, and a metal kerosene can in my hands. I can almost feel the gritty oil slopped on cold fingers. How I hated that chore! But things were to change dramatically in our rural area with the coming of electricity.

> "We make shadow figures on the wall, my grandson and I, and my mind drifts back . . . ."

Printed on yellowed paper the abstract to the family farm reads, 1938 —- REMC comes upon land. So few words, but packed with opportunities and power for the recipients.

Shiny new fixtures capable of invading even the darkest corners would replace the cumbersome lamps. The gas table lamp, pumped up by my father, then lit with dire warnings, had graced the dining room table. There, children had poured over lessons, played card games, and older folks read the Bible and newspapers, under its pale white glow. It would be retired to the attic. Overworked housewives suddenly would find their workload lighten. Flatirons need not be heated on a blazing range in hottest weather, an electric iron would glide smoothly over clothing which had been easily washed in a machine.

An electric sweeper would neatly and thoroughly clean corners and carpets, making housecleaning an enjoyable task.

Making seven-minute frosting and mashing potatoes would not be work for a strong armed person, as beaters did it easily.

As for storing food, no longer would blocks of ice be placed in an ice box, which occasionally overflowed the pan, calling for a good mopping. A refrigerator would serve well in its place.

An already weary farmer, on a hot summer day, with no breezes to turn the windmill, had sometimes pumped for hours, to supply water to the tank in the barnyard. What a blessing, to flip a switch, and have his work done!

The huge barn, mysterious with bins, rooms, stalls, mows and endless hiding places was formidable, unsafe by dim lantern light. How much pleasure now, for safe playing, and for doing barn chores quicker and more efficiently.

A young girl milking cows would no longer be surprised and frightened by a curious mouse, which might creep near her milking stool; now there was illuminating light.

Radio would become a companion of newspapers, in keeping the family informed of world events. Religious programs, farm broadcasts, music, and humorous shows would brighten the days and long winter nights.

Controlled heat, powered farm tools, fencing. . . the list goes on and on.

How inadequate those few words on the old abstract, and so incapable in relating the magnificent story of electricity — of an enlightened world — the day the lights came on for good.

What others said about Norris

"Dream on, you Senator from Nebraska, for your dreams mean but one thing. Your dreams, sir, mean that humanity may benefit, people may prosper, and human beings may be a bit happier.

— Hiram Johnson,
Remarks to the U.S. Senate
December 19, 1924

"When our commitment to the people is as complete as his, when in our several ways we work as hard and courageously as he worked, when we are as generous in victory as he was and as steadfast in defeat, then each of us it can be truly said that we have "served our country well."

— Senator Lister Hill, Alabama,
Honoring Norris May 16, 1961

"History asks 'Did the man have integrity?'
'Did the man have unselfishness?'
'Did the man have courage?'
'Did the man have consistency?'
And if the individual under the scrutiny of the historical microscope measured up to an affirmative answer to these questions, then history has set him down as great indeed in the pages of all the years to come.

There are few statesmen in America today who so definitely and clearly measure up to an affirmative answer to the four questions as does the senior Senator from Nebraska George W. Norris. In his case, history has already written the verdict."

— Franklin D. Roosevelt
Remarks to residents in McCook, Nebraska
September 29, 1932

"He was a Senator from Nebraska — but he belonged to the nation. He was, through most of his life, a Republican — but his statesmanship knew no party. He was a gentle, kindly soul — but he showed his enemies no mercy. He was devoted to serving the people — but his will, not theirs, cast his every vote. He became famous for the battles he lost — but he lived to see almost his every cause prevail."

— Theodore C. Sorenson
Special Counsel to President John F. Kennedy
Honoring Norris May 16, 1961

"We remember him not only for bringing actual light and power to the Nation's farms but also for bringing moral light and power to the Nation's deliberations."

— Hon. Arthur J. Goldberg,
Secretary of Labor
Honoring Norris May 16, 1961

"Senator Norris cared for the people. There was no arrogance about him. Usually he was forgiving; and always he was chivalrous and tender. He was characterized by tolerance and charity . . . the honesty of George William Norris is unassailable. His hand never touched a bribe . . . He did not use his privilege of patronage. Certainly he did not use it to his own advantage. He was a senator, not of Nebraska alone, but of America."

— Dr. Bryant Drake
Delivering Norris's funeral sermon
September 4, 1944

Bibliography

Books

Brown, Clayton. *Electricity For Rural America; The Fight for the REA.* Westport: Greenwood Press, 1980.

Childs, Marquis. *The Farmer Takes a Hand.* Garden City, N.Y.: Doubleday, 1953.

Clapp, Gordon. *The TVA, An Approach to the Development of a Region.* Chicago: University of Chicago Press, 1955.

Firth, Robert E. *Public Power in Nebraska: A Report on State Ownership.* Lincoln: University of Nebraska Press, 1962.

Garwood, John D. *The Rural Electrification Administration, An Evaluation.* Washington: American Enterprise Institute, 1963.

Hamilton, Carl. *In No Time At All.* Ames: Iowa State University, 1974.

Ickes, Harold LeClaire. *Back to Work.* New York: Macmillan Company, 1935.

King, Judson. *The Conservation Fight: From Theodore Roosevelt to the Tennessee Valley Authority.* Washington: Public Affairs Press, 1959.

Lasky, Victor. *Jimmy Carter: The Man & The Myth.* New York: Richard Marek Publishers, 1979.

Leuchtenburg, William E. *The Perils of Prosperity: 1914-1922.* Chicago: University of Chicago Press, 1958.

——, Franklin D. Roosevelt and the New Deal 1932-1940. New York: Harper and Row, 1963.

Lief, Alfred. *Democracy's Norris: The Biography of A Lonely Crusade.* New York: Stackpole Sons, 1939.

Lowitt, Richard. *George W. Norris: The Making of a Progressive: 1861-1912.* Syracuse: Syracuse University Press,

1963.

———, *George W. Norris: The Persistence of a Progressive: 1913-1945.* Urbana: University of Illinois Press, 1971.

———, *George W. Norris: The Persistence of a Progressive: 1913-1933.* Urbana: University of Illinois Press, 1971.

———, *George W. Norris: The Triumph of a Progressive, 1933-1944.* Urbana: University of Illinois Press. 1978.

Norris, George W. *Fighting Liberal: The Autobiography of George W. Norris.* New York: The Macmillan Company, 1945.

Norris from Nebraska, George W. Norris Foundation: McCook: Center for Great Plains Studies, 1991

Neuberger, Richard L. and Kahn, Stephen B. *Integrity: The Life of George W. Norris.* New York. The Vanguard Press. 1937.

Morris, Gene O. *Portraits Of The Past: McCook's First One Hundred Years.* McCook: High Plains Historical Society, 1982.

The Next Greatest Thing, Richard A. Pence, ed. National Rural Electric Cooperative Association: Washington, 1984.

Zucker, Norman L, *George W. Norris: Gentle Knight of American Democracy.* Urbana: University of Illinois Press, 1966.

Articles

Additional information obtained from the newspaper archives of *The McCook Daily Gazette* and *The McCook Tribune,* the *Red Willow County Gazette,* The *Omaha World-Herald,* and volumes of *The Congressional Record,* and *Rural Electric Nebraska* and *Nebraska Electric Farmer, Life Magazine,* and *Reader's Digest.*

Editor's acknowledgments

I thank members of the George W. Norris Foundation for their vision in the development in this book and their ongoing efforts to enlighten, to educate and to remind us about the remarkable man who led a remarkable effort to bring electricity to the rural countryside. Committee members include: Chairperson Flora Lundberg, Peter Graff, Ruth Leopold, Gene Morris and J.T. Harris, Jr.

Linda Booth Hein, Site Supervisor at the Sen. George Norris State Historic Site in McCook, was most helpful in directing research, loaning materials, offering advice and supporting this project as was Cloyd Clark, an advocate in all matters pertaining to George W. Norris. The staff at the McCook Public Library offered valuable assistance in locating research materials. Three instructors at McCook Community College provided the initial energy for the book when they judged these stories in essay form. Those instructors include: Susan Watts, Chet DeVaughn, and Rod Horst.

It's been 70 years since the spirit of cooperation helped form so many successful rural electric cooperatives and that neighborly spirit of cooperation is very much alive as evidenced in the support given to this project. I am grateful to Jim Phinney at McCook Public Power District for his help in locating photographs from their archives and to Phyllis Marsh for loaning a valuable photograph and in her support of the George Norris legacy.

Renee A. Butler of the National Rural Electric Cooperative Association was valuable in helping secure photographs for this book and putting me in touch with Ben Weybright at Asman Custom Photo Service.

Lori Cobb and Lorri Sughroue helped typeset and proofread the original letters and the manuscript. For their assistance, I am forever thankful as well as to Jack Rogers.

Members of the Inland Sea Writers should know their suggestions and input regarding this book and other writing is always appreciated. Thank you, Merrill Ream, Bryan Jones, Mike Pruter, Dale Marie Bryan, Ginny Odenbach, Walt Sehnert and Amy Strauch.

— B.L.C.

PHOTO CREDITS

The bottom photograph on Page 163 appears courtesy of Phyllis Marsh.

All of the other photographs including the front and back covers appear courtesy of the National Rural Electric Cooperative Association from photographs contained in the publication "The Next Greatest Thing."

Circle of Brightness Order Form

Use this convenient order form to order additional copies
of
Circle of Brightness

Please Print:

Name_____

Address_____

City_____ **State**_____

Zip_____

Phone(   )_____

_____ copies of book @ $19.95 each $ _____
Postage and handling @ $ 2.50 per book $ _____
NE residents add 5.5% tax $ _____
Total amount enclosed $ _____

Make checks payable to George W. Norris Foundation

Send to George W. Norris Foundation
P. O. Box 884 • McCook • NE • 69001